Turbulence
and Disorder
in World Systems

Other books by John J. Moran

Prisoner Of Conscience: 2014

Know Your World: A Geographer's Guide To The Anthropocene Age : 2016

Applied Geography: The Formulation Of A New Ecological Science: 2018

Turbulence and Disorder in World Systems

Cognitive Immunological Reaction

OR

Increasing Entropy

John J. Moran FRGS

Book cover:
After an illustration in Time Magazine: 13 February 2018
W. J. Hennigan: 'The New Nuclear Poker Game': 'Saucer-shaped craters in Nevada, resulting from underground nuclear tests during the Cold War'.

Contents

Illustrations

Acknowledgements

Having spent a lifetime in studying and teaching geography, I owe a debt of immeasurable gratitude to all who have contributed to opening my mind to contemplate the wonder of being alive in the fascinating environment of our Earth-world. Foremost, are all the members of my family, my ancestors- who crossed the seas and arrived at the Atlantic fringe of Europe thousands of years ago, bequeathing me their name (*Moran= the people of the sea),* as well as all my earthly kin who have given me so much nurture and care. My beloved late wife, Mary, who always found time to listen to me rehearsing the next day's lesson or lecture, often tempering my thoughts by the spiritual quality of her outlook, did much to focus my attention on the inequality which besmirches our world. Her empathetic influence expressed in gentle words of advice such as, *'you have to walk a mile in their shoes to know how they feel or act',* will remain forever engraved on my mind. Then, our 7 children have kept us on our toes, keeping up with their interesting explorations and journeys through Life. John (Jnr.), Un. of Queensland, who has conducted language courses and workshops for teachers and professional educators in Oman, Dubai, Brunei, China, The Philippines, Viet-Nam, and is linguistic consultant on migrant entry to the Australian government, has enabled me to share in his experiences and to learn so much about the geographical connections of world systems, that I feel as if I have carried out fieldwork whilst sitting in my armchair. Andrew, working in personnel management for Australian Telecom, keeps me informed about life in Melbourne, Bernard, qualified as an Environmental Scientist, keeps me fed with scientific developments, and David, whose linguistic studies have taken him to Berkeley (Un. of California), Peru (Arequipa Un.), and Moscow Un., and is now working as a Librarian, has provided me with a valuable insight into South

American and Russian political and economic affairs, as systems of human organization. Our three daughters, have also become intrepid explorers, as well as contributing to the well-being of the world about them: Kathleen, having specialized in remedial teaching, has revealed for me how dysfunctional systems of human organization may be connected with the problems arising from increasing disadvantage and impoverishment, which are dividing society into the haves and the have nots; Anne, qualified both as a Nurse and Teacher, took us to Indonesia in mind during her voluntary work there, and continues to keep me informed through her family travels; Margaret, went to Australia to gain experience as a nurse, and has settled as a District Nurse in Melbourne. My grandchildren have followed suit and continue to enrich my view of the world by their studies, interests, and travels, as well as in their capacity to share their enthusiasm with me. They offer me a new intellectual lease of Life.

I owe so much to St. Mary's College of Education, Strawberry Hill (St. Mary's University), where my mind was opened to the Principles and Practice of Education, and to Birkbeck College (London University), where my foundation perspective of the connections between geophysical systems and human organization in our world was formed, that it would take a volume to express my gratitude. In particular, I recall the impact of the Rev Dr. Dunning's lectures on the Etymology and Semantics of Language, as among the most interesting and thought-provoking of all my experiences: he was able to pass on the inspiration he had received from his research tutor- the great teacher C.S. Lewis, to his own students. Equally, I hold in grateful memory, Professor Eva Taylor, Ph.D... DSc., who gave her students at Birkbeck College (London Un.,) so much inspiration. She was the first woman to be awarded a doctorate in London University, and I believe the first woman academic in the UK, to achieve a well-deserved double doctorate. Also, I realize how privileged I was, to have had Sir Peter Hall, the pioneer geographer of Urban design and Planning, as my research tutor during my studies at London University. I am greatly indebted to my daughter Kathleen Percy and son-in-law Tom Percy-Biologists and Educationists for taking on the tedious task of reading the draft of this book, and pointing out obscurities and non-sequiturs in meaning, Equally, my daughter Anne Prior and son-in-law Martin Prior-Geophysicist and SR Nurse, have provided valuable information concerning frequency and severity of epidemics: Martin's guidance through the baffling chain of interactions relating to vortical entanglements in climatic *'Energy Fields of Flow'*, still keeps me awake at night. Also, my son Bernard has given me much food for thought, relating to his research at Cranfield University on *'Satellite Imaging'*, concerning the impact of human activities upon habitats on Earth. My children in Australia have also contributed valuable insights and information about changes in geographical patterns, linking Australia to developments in the Asiatic Region: John (Jnr.), Lecturer in *'Language & Education'* (Queensland Un.); Andrew, *'HR Management; Australia Telecom'*, covering USA, India, & UK.; and Margaret, District Nurse in Melbourne. In addition, my grandchildren, are always ready to come to my aid, when the computer goes against me. Any errors are my own.

John J. Moran

Introduction

*('When our last hour comes, we shall have the great and ineffable joy of seeing the **One** whom we could only glimpse in our work': Carl Friedreich Gauss: 1777-1855: 'The greatest Mathematician since antiquity').*

('Admitting uncertainty not only bridges the gap between science and religion, but also can do the same, when applied to Life's seemingly perpetual cycles of disputes': Richard Feynman).

Geography is a custodial science, its discipline focused on keeping track of the complex web of interactions, linking human activities and their impact upon the potentially violent reactive geophysical systems which account for, or have contributed to critical environmental change, such as *'The Climate Warming Crisis'*. In that respect, it compares with Theological principles and teaching, for both converge in concern for the well-being of people (and all living things), as well as on keeping the world in good shape for the benefit of future generations. Whereas, *'Theological and Theosophical philosophy'*, may be understood as energized mainly by *'Spiritual Logic of Mind'*, geographical analysis comprises both *'pragmatic logic'* (Ecologically based decisions about the exploitation of resources, so as to minimise the likely detrimental impact upon the environment), as well as managing human resources sensibly, inspired by an unselfish *'spiritual logic of mind', which* involves maintaining a caring outlook for the least able members of society, comprising the elderly, children living in conditions of dire poverty, as well as the disabled and mentally impaired.

1

The Challenge: Economic gain & discord in society relationships; or Fair shares &harmony in systems of human organization: Many of the problems facing our modern world have arisen through greed and the inordinate acquisition of wealth by a few, at the expense of the working fraction of the population. Thus, the store of resentment resulting from the harsh and cruel treatment of people condemned to lives of *'slave labour'*, in the long distant past, continues to smoulder, with slogans, such as *''Black lives matter'*, which can easily flare into episodes of vicious mob violence. People living side by side in peace for ages, suddenly become maddened by hate, prepared to kill each other, and even to destroy the very environmental infrastructure upon which their well-being depended. For example, in1993 the ancient *'Mostar Bridge'* in Sarajevo, built in the 16th Century, was destroyed when war broke out between the Christian Serbs & Croats and the Bosnian Muslims.

It had stood for 427 years, giving free access to both sides of the city, enabling Muslims and Christians to mingle, intermarry and live with mutual respect for each other, until inflamed with hate, goaded by a revival of hazy memories of past hideous conflicts, dating back to the 13th Century under the cruel tyrannical rule of Osman 1st, who established the Ottoman Empire which eventually extended from the Danube to the Nile, leaving a trail of havoc and fear in its wake.

The Bosnian V Croat-Serb War: Store of good and bad in the human heart and mind: Clashes between these factions, began in 1992, after the collapse of *'The Socialist Republic of Yugoslavia'*, at the end of *'The Cold War'*, and escalated into a bitter conflict when the USA and the EEC (European Commission), both announced their recognition of Bosnia and

Herzegovina as prospective members of the developing European Union, in April 1992. However, it may be seen as a flare up of deeply rooted hateful memories, harking back to the genocidal cruelties inflicted by barbaric tribes, as invaders from the desert regions around the Caspian and Aral Seas, which remained smouldering from the long distant past. Writing in 1844, Alexander Kinglake refers to the pyramids of skulls, left in cities including Belgrade, as a warning to any thinking of resistance. *('Eothen': Alexander Kinglake')*. From the 14th to the 18th centuries, the Ottoman reign of terror prevailed, extending from the Danube to the Saharan Desert, imposed by swaggering Janissaries strolling the streets with their long, curved swords, ready to strike off the head of anyone they chose. They were trained as special Turkish troops, and the Sultan's Praetorian guard, formed from tributary Christian children. That is how a store of evil can accumulate in the human heart, smouldering on, with the probability of blazing into even more catastrophic episodes of traumatic disorder, such as that imposed by the Nazi SS Brigades in the 2nd W. War, and in the renewed bitter conflicts breaking out across the Arabian world today.

The baleful influence of these antagonistic memories continued to have a significant effect in the preparations for the 1st World War, when Germany in alliance with the Ottoman Turks, promised to restore the Ottoman Empire, in exchange for their assistance in seizing control of the oil resources in the Middle East. This alliance was renewed in the 2nd World War, resulting in the hideous loss of life during *'The War of the Desert Rats' (7th Armoured Brigade),* against the *'Africa Korps'* under the general ship of Erwin Rommel, which raged across N. Africa to Egypt. The wounds inflicted across the Middle Eastern Region of the world at that low point in human affairs, continue to fester on, causing millions to flee in desperation, for fear of their lives, in search of the optimum location to ensure a better and safer life. Traumatic experience spawns bad thoughts, as well as a callous attitude towards the sufferings of others. In turn, indifference breeds contempt for human life, and a readiness to inflict cruelty upon others without compunction, paving the way to extreme cultural decline, leading to chaotic disorder in systems of human organization and government. This reprehensible pattern of behaviour may be observed amongst the horde of migrants, fleeing from war-torn zones of N. Africa and the Middle East. For example, when a BBC team of reporters came upon an elderly woman, clad in a tattered burka, shoeless with lacerated feet, in tears and struggling to carry a young child, alone along a rough stony track, they took pity on her and offered her a lift. Eventually, some miles further on they caught up with a group of her tribal menfolk, who had gone before, leaving her behind. Instead of welcoming her, they berated her for accepting help from strange men, without their permission, forcing her to get out and struggle on for many more miles, without lifting a finger to help her. *(BBC. World News: 2020)*. We may learn from this train of events that all we come to know and perceive bears the marks of being shaped by previous attitudes, behaviour, values and associated conditions. That is how **wicked behaviour and experience** may grow like a spiritual cankerous tumour, to contaminate social relationships and create unmanageable disorder in systems of organization at world level, **unless held in check by the healing power of brotherly love.** *('If I am without love, it will do me no good whatever. Love is always patient and kind; it is never jealous; love is never boastful or conceited; it is never rude or selfish; it does not take offence and is not resentful. Love takes no pleasure in other people's sins but delights in the truth. It is always* *ready to excuse, to trust, to hope, and endure whatever comes. Love does not come to an end. But, if there are gifts of prophecy, the time will come when they must fail; <u>or the gift of languages, it will not continue for ever; and knowledge—for this, too, the time will come, when it must fail. For our knowledge is imperfect, and our prophesying is imperfect: but once perfection comes, all imperfect things will disappear')</u>;St. Paul:1 Corinthians13. 3-10*

How is it possible, that an abstract spiritual sense of mind could exert a powerful healing effect? Medical scientists have noted *'A Placebo Effect'*, by which the mere thought of *'being cared for'*, appears to have a similar therapeutic curing quality to known pharmaceutical medication.

With that knowledge, just consider what could have happened, if the decision had been made **to cancel all national debts,** at the meeting of the *'G 7 World Council'*, *convened* to discuss a strategy for economic recovery, after the disastrous collapse of national economies across the world, as a result of *'Covid Pandemic'* debts. The probable autocatalytic effect would have been to unite all nations as one, knowing, ***'we are all in it together'***, bringing relief and hope especially to the poorest nations, burdened with unmanageable debt. *(G7 Conference: Cornwall, UK: June 2021)*.

The material world delusion: Reactions arising from the *'Covid Pandemic'* crisis serve as a warning that the time has come to take a clear look at the causes of turbulent disorder, becoming even more menacing to the welfare of our world. The scale of impending world disorder we are facing cannot be exaggerated, compounded as it is of problems arising from environmental pollution with those which are associated with extreme inequality: e.g., Millions of refugees in desperate flight, for fear of their lives, or to escape from conditions of dire poverty. But not all the scientific community appear to share this concern. Thus, vast resources continue to be frittered away on useless speculative ventures, such as costly experiments in sending probes to Mars, whilst thousands are dying from pandemic disease, and lack of basic needs- **such as clean water to drink, down here on Earth: e.g.** *('Astronomers have discovered the first signs of light to appear after "the Big Bang" which brought the Universe into being over 13 billion years ago'): BBC. World News: 25 June, 2021).* Thus, hordes of highly remunerated physicists appear to be obsessed with what could be happening, at the *'Cosmic and Astronomic'* scale of *'Time & Space',* whilst remaining unable to see what is happening beyond their nose, here on Earth, where thousands of people are losing their lives, and money is losing value and becoming worthless.

Valid knowledge garbled to become misinformation: Scientists who now claim to be able to see the first *'Light of Creation',* which has been travelling towards them for *'13,000,000,000 Light Years',* seem to have forgotten the knowledge gained by the great scientific minds upon whom they stand: ***('Isolated particles are abstractions, their properties being definable and observable only through their interactions with other systems—what you think is real is not real after all'):*** *Niels Bohr:1885-1962: Pioneer Physicist & Philosopher of Science: Nobel Prize, 1922: 'Atomic Spin Model', linking electron energy + magnetic radiation: Explaining the visible spectral patterns of the hydrogen atom.* Niels Bohr, narrowly escaped arrest by the Nazis, suspected of slowing work on their atom bomb, and became Britain's representative in the Manhattan Project, which shortened the war, saving millions of lives. Now all scientists agree that nothing in the Universe can be defined as solid, or material. All things, as macro-molecular assemblies, are fizzing in a state of change. If nothing can be considered as solid on Earth, it would be folly to waste valuable resources in experiments '*beyond the ends of our Earth-world & in conditions of space and time'*, that no longer exist.

('Can everything be explained in terms of the behaviour of the particles they are made of?'): *George Ellis: co/author with Stephen Hawking: 'The Large-scale Structure of Space-Time':1973).*

('Is there a theory of quantum gravity to be discovered which may reconcile the mystery of the information field?': **& 'Can infinite space-time explain our exceptionally well-ordered Universe?':** *Joseph Silk: Director of Philosophy of Cosmology: Oxford: NS: 'Cosmic Conundrums':8 March 2014.* **(116).**

Whilst these hypothetical questions, based on assumptions and speculations about the mystery of '*Creation, and how all things came into existence'*, are of interest, there are more pressing priorities here on Earth, requiring the leadership of Geography, **'The Queen of Sciences', to keep minds focused on what really matters and demands our concern, the deterioration of living conditions in our home-world, and the eradication of poverty.**

Mystery of the 'Information Field': 1) Deliberations on Metaphysics, the ultimate science of 'being and knowing' *(Plato-> Galileo-> Gauss->Maxwell->Darwin->Einstein->Bohr-- ->Rees-> Penrose),* indicate that all the connections perceived to be linking the interactions between the attributes of our Earth-world and systems of human organization, can be understood and explained in terms of the transmission of

7

power along energy-gradients: **pressure, temperature, concentration, force, alkalinity, etc.):**

2) The flow of energy appears to be regulated in some way by *'wave frequency' (high frequency- strong flow-> onset of turbulence+ competing rhythms+ i cacophony-> chaotic patterns of disorder):*

3) Earth perceived as a spherical body, rotating & spinning round the sun *(1 rotation @ 1,000 mph =1 daytime):*

4) Advent of the Moon to form a binary system with Earth: *(4.5 billion years ago-> acted as a brake to reduce previous speed of rotation of about 25,000 mph, providing 'gravity bound' conditions favourable to emergence of Life):*

5) Discovery of the forces of 'gravity' & 'electro-magnetic radiation' as related fields of transmission of 'energy/light/heat/nuclear forces of attraction and radiation*:* All these energy flows can now be understood as generated by the effects of vortical motion, which releases radiation in the form of kinetic heat from spinning bodies: *('Electrons orbiting in atomic shells; molecular systems; gravitational attraction; planets orbiting stars; galactic systems, e.g. 'The Milky Way'; etc.).* In this respect, the heat energy emitted by the fluxing changes within, and between our Earth-world's structural shells, including the large scale *'Systems of Circulation'* (Oceans; Biosphere; Atmosphere; Stratosphere; +'The protective shields'-Ozonosphere & Magnetosphere), must be taken into account.

6) Quantum Fields of Vorticity: No one incident of vortical motion can be studied without taking into account the effects of other vortical phenomena with which it is entangled. For example, air warmed by the *'North Atlantic Drift'* ocean current rises to form cyclonic *'Low pressure systems'*, which rotate anticlockwise in the N. hemisphere mid-latitudes, because a rotating vortex over the Earth's *'initial rotating frame'* is deflected to the right of its direction of motion. Similarly, rising warm air *'Lows'* would be deflected to flow clockwise in the S. Hemisphere. In the same way, the rotating Sun exerts a vortical influence, as *'an initial rotating frame'*, over all the attributes of Planet Earth. Since our Sun is but one constituent star of ''*The Milky Way*'' Galaxy, we are also caught up in vortices of infinitesimal magnitude. The immense flow of energy from vortical systems does not lend credence to the concept of *'Increasing Entropy'*, with inevitable dissipation of energy in the *'Universe' and the obliteration of all things by the flailing swords of Shiva'*. On the contrary our Earth world appears to be able to store energy, like a battery. Problems of mounting concern, such as *'climate warming'*, have more to do with runaway energized systems, than the run-down of energy: cf.: - *'The Lorenz Waterwheel Experiment'* & *'Chaos Theory'*: David Lorenz1961.

7) Mystery of *'The Goldilocks's Enigma'*: With the knowledge we possess, confirming past wild and unpredictable behaviour in Earth's geophysical systems, it would not appear to be a propitious place for Life to emerge and develop. In fact, there have been 3 major extinction events during the last 270 million years. Moreover, although the Sun has become much warmer since the inception of life, Earth seems to have managed to maintain a state of homeostatic balance, in support of Life processes. Somehow, the Earth has managed to remain a remarkably auspicious location for the support of Life, despite the high probability of turbulent events, leading to chaotic disorder.

9

8) 'The Maxwell Conundrum': When Maxwell saw the connection between photons of light and the transmission of electromagnetic power as a network of interactions, linking the emergent properties of *'atomic/molecular systems'* with the parameters of *'geophysical systems'*; all fusing into *'patterns of ambient order'*, it enabled him to comprehend the mystery of *'Life coming into existence'*, in a world environment apparently prepared to receive and nurture it, as an ***'immaterial reality of entanglements'***, including himself as part of it, as well as the unfolding knowledge in his own mind. Thus, he may have realized that the flow of *'electromagnetic power'*, due to changes in the magnetism produced by an electric current, and related to the release of energy by electrons spinning in atomic shells, could be understood as regulated in a strangely unexpected way. Instead of a probable exponential increase in wave patterns of transmission, his insight opened the probability of a reciprocal reaction in the flow of energy he was observing, whereby the heating effect could be tempered by a regulating cooling effect: (*'Maxwell's Demon'*: Lord Kelvin; 1894). In this context, atomic orbiting shells could be regarded as *'sinuous vortices'*, acting as a regulating influence.

9) All patterns of flow, in *'Earth's Large Scale Patterns of Circulation'*, must be considered as much more complex than, 2-dimensional streams of flow: e.g. *'The Jet Stream'*; *'North Atlantic Drift'*; *'Continental Drift'*; *'Mid-Ocean Rifts'*; *'Atmospheric Shells'*; *'Ozonosphere'*; *'Magnetosphere'*, but as enmeshed vortices, acting according **to an innate principle of order.**

Crisis of climate warming: Cooling effect of enmeshed vortical reactions: The *'Jet Stream'*, viewed as a corkscrewing vortex, following a spiral path turning in the same direction as the '*high- and low-pressure systems*', which it is steering, may be understood as a subsystem of adjustment, and not merely a driving force. In the same way as Maxwell opined that the '*magnetic field*' produced by an '*electric current*', could be tempered by the non-linear cooling effect of variations in the strength of the current, think of the '*Jet Stream*', as a '*vortical system*', whiplashing and jerking in such an unpredictable way, that resulting synoptic weather patterns may vary widely, ranging from raging destructive hurricanes, to relatively gentle balmy '*soft regimes*', such as experienced in S W. Ireland and the sub-tropical gardens of Inverere.

Note:- * In vorticose motion the sinuosity in centripetal patterns of flow, varies in proportion to the systemic energy. It follows, that thoughtless activities such as '*Formula 1 Racing*', & '*Space Travel Rockets*', travelling at 37,000 m p h.., can be expected to have a catastrophic effect upon the tenuous gaseous shells of our Earth world, especially the fragile *Biosphere* and our vitally important protective shields, '*The Ozonosphere & Magnetosphere*'. Earth's curvature is constantly changing, so it is a total mystery that its environmental parameters should be so supportive to Life processes. The Earth is not spherical, but irregular in shape, flattened like a doughnut, '*geoidal*', and constantly flexing and bulging as it twists and jerks on its way, along a twisting and spasmodically variable vortical path. The fact that it has remained in a dynamically stable condition, to support '*Life*", against all the geophysical odds, is truly astonishing, and beyond comprehension. There must therefore be an internal system of regulation, empowered with knowledge to counteract external destructive events. That presents indisputable proof of **an inherent principle of order,** within the '*Geographical Life/Earth world System*', from the moment of inception:(*'Just 6 Numbers': Martin Rees: Astronomer Royal*). * Thus, it becomes a matter of urgency to investigate the destructive effects of foolish human activities, posing the highly probable risk of inflicting irreparable damage upon Earth's delicately balanced *'Large Scale Systems of Circulation'*, because of induced vortiginous disturbance: e.g. *'Formula 1 Racing/Gambling'-> pollution+ hypoxia; 'Space Travel Rockets'-> pollution by accumulation of debris+ destruction of the protective Ozonosphere.*

* It would make immediate good sense if traffic were directed to drive on the left in the N. hemisphere, since vortical motion gives rise to an opposite reaction, so a high volume of traffic driving on the right, tends to trigger anticlockwise circulation, rising air-*'Low Barometric Pressure),* with increasing velocity of flow, to develop. Traffic driving on the left would tend to counteract the formation of highly destructive *'Hurricane Whirl-winds'.* That would save several billion dollars per year in the USA, by reducing the vortical spin of destructive hurricanes experienced regularly, by heavy traffic driving on the right.

* It would make longer term, ecologically wise sense, to pay heed to Maxwell's intuitive insight that, **a strongly destructive vorticose wave pattern of circulation could be modified by an opposite and weaker vorticose reaction, triggered by the main vortical thrust, but taking a longer path, with a braking effect (slowing & cooling), acting as a regulatory system of adjustment and control.** For example, instead of dwelling on the naïve concept of 'The Carbon Footprint', as a method of counteracting the hazards of a *'Warming Climate',* it would make better use of the valid scientific knowledge gained, to concentrate on *'The Large Scale Patterns of Earth's Entangled Circulatory Systems',* in

order to investigate their interactions and take measures to maintain our precious *'Biosphere'* in the extraordinary dynamically stable state, upon which all living things depend. *(cf. Edward Lorenz: 'Large Scale Motions in the Circulation of the Atmosphere': MIT.*

* Although it is known that the oceans contain about 50 times more CO^2 than the atmosphere and about 10 times more than that held in plant and soil carbon stores, it is not the most toxic polluting *'greenhouse gas'*, assumed to present the greatest hazard as a result of a warming climate: *(cf. IPPC: Intergovernmental Panel on Climate Change: Nov. 2015)*. The oceans, seas and the vast areas of thawing *'Tundra'*, also contain enormous stores of Methane, CH4, held as a effect of the thermal gradient. The holding capacity of both gases is temperature dependent, so most of these stores are held in deep cold waters, especially in fossil salty seawater held under less dense fresher water, like the Arctic Ocean and Baltic Sea, as well as the fossil ocean recently discovered beneath the vast Antarctic Ice Sheet. Methane is 23 times more polluting than Carbon dioxide, and is already seeping out in vast quantities, beyond any possibility of control, from the thawing *'Tundra'*. However, ***'Maxwell's Conundrum'*** may show how to make the best use of knowledge we possess.

* Since the rate of accretion in heat energy can be moderated by the regulatory power of vortical motion, it should be possible to mitigate the harmful effects of *'climate warming'*, by extracting heat from warming ocean surface circulatory currents, using deep colder water as a cooling agent. Already, several *'Ocean Thermal Energy Conversion Plants'* are in process of development, and the ensuing possibility of unlimited cheap energy, freely available to the whole world, is

within sight. *(cf. NS: Helen Knight: 1 March 2014: 2958: p.50:'Sea Change: Ocean Power'): (cf. Moran, J.J: 'Know Your World: Ch.5: 'Mysterious Water':pp.178-212: 2016).* In this context, *'Geothermal Energy'* is known to be available in abundance. For example, in the Roman Baths of *'Aquae Sulis',* hot chalybeate springs are still flowing at a surface temperature of 46° C. If tapped at a depth of a few kilometres, who knows how much cheap electrical power could be unleashed? *(cf. BBC News: 10 July 2021: 'Development of Geothermal Energy in Cornwall')'.* If all these developments come to fruition, there would be no need for more Nuclear Power Plants, with all the hazards they entail, such as the difficulty in disposing of lethal waste material, as well as the high probability of deadly nuclear accidental emissions. Furthermore, it would eliminate the even more traumatic probability of hazards, arising from all the amateurish experiments known to be taking place in back-room laboratories, ill-equipped for the danger of handling *'plasma'* - a substance not known on Earth, at the infernal temperature of 800° C, comparable with the surface of the Sun. These experiments apparently rely upon the force of *'electro-magnetism',* to contain the all-consuming heat of *'plasma',* but what would be the consequence, if 'Earth's polarity were to flip, as it has done many times in geological history? What exactly is the connection between *'Radio waves' & Electro-magnetic energy'?* In respect of this uncertainty, it is time to be aware that it becomes foolhardy to take the dynamically stable state of planet Earth supporting *'Life processes'* for granted, as robust and permanent. On the contrary, geophysical systems must be understood as constantly fluxing fluid systems, easily tipped into a state of imbalance as *'Life support systems',* by violent disturbance such as the onslaught of space rockets travelling with immense vortical power, capable of tearing a hole through the precious shield of the *'Ozonosphere'*. Without that protective shield, Life on Earth could not exist.

The key to mitigating the destructive effects of *'Climate warming'***:** *('In fluid dynamics, we can summarize the physical state of affairs by means of a field of flow. In general, the field of flow is a vector field. We associate a vector quantity with each point in space, namely the flow velocity v at that point. For a steady flow the field of flow is stationary --The field gives the properties of the space from which we deduce the behaviour of particles in that space. If the flow is irrotational,* (i.e., not rotating), *as well as steady, we call it 'potential flow'-- For a 'uniform rotation'* (i.e., Planet Earth), *v is proportional to r: In the field of flow of a vortex v is proportional to 1/r'):* Professor A.D. Moore (Un. of Michigan): 'Fields of Flow'. **(117).**

The diagrams depicting *'variations of fluid velocity from the centre',* for both a *'Uniform rotational field of flow',* and a *'vortical field of flow', show that there are 2 equivalent ways of deriving the relative magnitudes of the flow velocities from such fields of flow: (a) from the widths of the tubes of flow, and (b) from the distances between lines of equal velocity potential':* Prof. A. D. Moore: 'Fields of Flow': pp. 453- 456.

Entangled vortical fields in Earth's Largescale Circulatory Systems of Flow: Dynamic Earth, as we know it from geographical experience must be understood and respected as an immensely energized *'field of flow',* far beyond any possibility of management or control by any human system of organization which may be devised, with the best will in the world. But, as we have learned from the naïve wrangling observed during the last *'United Nations Summit Conference'* on climate change (Paris 2015), we cannot rely upon the best will in our *'world-full of error'*. It becomes the duty of the discipline of Geography, charged with monitoring the interplay between systems of human organization and their effect upon geophysical systems energized at an infinite scale of magnitude, to steer the way out of this obscuring mindset, rooted in the past, with such misleading slogans as, **'*Our share of the carbon footprint*' & '*Black lives matter*'.** All lives matter, and there is no such thing as the carbon footprint. All have benefited from burning fossil fuels in the past, in the form of infrastructure now taken for granted, such as railways and means of travel. Now is the time to think and act in the light of **'Logic of Mind', refined by the spiritual energy of 'Deep Ecology'.**

Symptoms of bitter resentment, lingering from past sufferings and wrongdoings, which continue to flare up into destructive conflicts, such as we are witnessing throughout the world today, could be easily resolved by a change of heart at world summit level. Instead of apportioning debts incurred by the costs entailed in countering the menacing effects of *'Covid Pandemic & Climate warming',* all debts could be cancelled, and the necessary vaccines could be made freely available to the poorest nations by the richest, since *'no one is safe, until all are safe'.* Also, there is urgent need for action to eradicate the pernicious effects of large-scale gambling, promoted by ruthless syndicates as *'an opiate to fleece the people':* e.g., *'Formula 1 Racing'; 'Football Club tycoons'; 'Lotteries',* etc. In that way, the desperately poor and impoverished elderly fall easy prey to wicked greed, as well as the younger generation, by the seductive lure of becoming rich. The poorest, especially children living in extremely impoverished conditions, are particularly vulnerable, easily addicted to gambling as a way of escape from *'rags to riches.* If only the essential lessons, we have learned from geographical experience were to be applied with common sense, cankerous discord could be healed, giving way to harmonic peace and well-being. **Love and care are indeed closely entwined, as a charismatic principle of order, with a curative effect.**

Scale and vagarious complexity of geophysical vortical fields of flow: Problems arising from climate change are entirely different in scale and complexity, from those associated with turbulence in systems of human organization and must be understood as *'variable energy fields of flow'*, in which energy builds up, and can be stored as in a battery, just like the accretion of energy in a macro-molecular living organism. In this way, our planet Earth-world is constantly changing in shape and momentum, varying in *'velocity, strength, and continuity'*, as depicted in the diagrams illustrating *('2 dimensional fields of flow, comparing a Uniform rotational field of flow with a Vortical field of flow')*: W. E. Rogers: *'Introduction to Electric Fields': pp. 454)*. **(117).**

We may learn from the diagram depicting a 2 dimensional pattern of flow in in the *'vector fields of vortical systems'*, such as the energy transfers in *'Earth's geophysical circulatory systems; the electrostatic field; the magnetic field; and the field of flow for an electric current'*, **that the key to making wise decisions in order to mitigate the deleterious effects of climate warming lies in understanding the relative magnitudes of the fields of flow.** Consider, for example, the alternating pattern of *'High- & Low-Pressure barometric systems'*, which define the everchanging synoptic weather conditions experienced in the British Isles and Western Europe, as successive 3-dimensional vortical systems of energy flow, carrying warmth from the Gulf of Mexico via the 'N. Atlantic Drift Ocean current to W. Europe, and as far as Scandinavia. Note that a high-pressure system (descending cooling air), may be understood as a *'3 dimensional vortical field of flow, varying in fluid velocity from the centre',* pressing down on the Earth's surface. The direction of flow in high pressure systems (anticyclones), in the mid latitudes of the N. Hemisphere, will be clockwise, because any moving body – over the *'initial rotating frame of the Earth',* is drawn to the right in the N. Hemisphere: (*cf. The NW. & SE. Trade Winds blowing towards the Equator).*

Similarly, low pressure systems (cyclones) - warmer air rising, are *'energized fields of flow',* weakening in velocity as buoyant air rises, and cooling as *'potential energy'* is dissipated through gathering water by phase-change to vapour. In this way, air warmed over warm ocean currents- must also be understood as enmeshed vortical fields of flow, able to pick up millions of tons of water, with the potential of disposing of it as heavy rainfall by condensation, which in turn releases latent energy, as well as kinetic energy on descent to the Earth's surface. In this way, warm katabatic winds develop on the lee side of mountain ranges, like the Alps, and Rocky Mountains of N. America. Known as *'fohn' winds,* they are welcomed as a climatic bonus, providing early spring growth to crops. Thus, there is a continuous cycle of exchange in vortical fields of energy flow, whereby, warm rising spiralling flow gives way to cooling descending vorticose flow, switching in direction of rotation at each turn of event, following an octal pattern of **dynamic change in the velocity of the field of flow.** Another way of visualizing this dynamic pattern of vorticose entanglement, would be in terms of increasing and decreasing sinuosity of flow, in inverse proportion to energy/velocity of flow. Just as a river, considered as a vorticose sinuosity of flow, can be understood as **a 'system of fluid velocity', waxing and waning in** inverse proportion to fluctuations in magnitude of energy, in conjunction with variations in velocity, strength and continuity in a 3-dimensional field of flow, so *'climate warming'* presents a far more complex problem than notional *'carbon footprints'*.

Earth's vulnerable atmospheric shells:

85- 600 km. Thermosphere: Ultraviolet radiation; gamma rays; X rays; ionization of atoms & molecules-->gases --> Temperature raised by several hundred degrees

50- 80 km. Mesosphere: Quasi-Biennial Oscillations (Oxford). Little known; 'Lightning sprites'

15- 30 km. Ozonosphere: O^3 content less than 10 pp million; Tropics-peak at 20-30 km; average atmospheric content 3 ppm; Towards Poles- 12=20 km. Vital protective shield, easily destroyed by CFC gases: Life could not exist without it.

14.5-50 m. Stratosphere: Pioneer Meteorologists (E. Lorenz et al.)->holds vital clues to synoptic weather patterns. Extreme Cumulonimbus anvil thunder clouds reach up to 21km

Earth surface-14.5 km. Troposphere: Synoptic weather--> interaction of geophysical systems with Life Processes & effects of human systems of behaviour and organization. Breathable atmosphere and biosphere limited to 3 km.,(10,000 ft. Extremely fragile).

Evidence of Earth's interior warming: There are clear signs of strengthening fields of flow in the vortical enmeshments of Earth's interior structural shells: viz. 1) Thinning of *'Mantle'* under Greenland by about 80 km: i.e. melting from below; 2) Extraordinary rapid melting of glaciers and *.Ice sheets'* in both hemispheres; 3) Recent discovery of emissions of *'rare gas isotope Helium 4',* known to be released by the nuclear decay of *'Uranium & Thorium'*, at a temperature of over 400C, probably triggered by repeated impacts of immense explosive force, such as continuous experiments with nuclear warheads since 1945.

Facing the facts: The last *'Global Conference on Climate Warming'*, (Paris 2015), became a shambles, like children squabbling over sharing a bag of sweets: *'The Carbon footprint euphemism'*. There is no time to waste in repeating such a charade. The time has come to pull together, if there should be any possibility of making good use of the valid knowledge we have gained.

Some useful managerial tips:

1) **All traffic in the N. Hemisphere should be directed to drive on the left:** This generates a counter, descending *'Vortical Field of Flow'*, with the effect of damping down the effects of strongly rising *'Vortical Fields of Flow'*, which cause *'Hurricanes'* to develop, as whirlwinds of highly destructive force. This would save the USA., billions of dollars a year. Curiously, by adhering to the British custom of driving on the left, **Australians have done the right thing to suit** *'xerothermic'* conditions, experienced in Australia's hot dry climate. In the S. Hemisphere, a clockwise *'vortical flow'* becomes an ascending *'Low Pressure System'*, able to carry fine desert dust high into the *'Stratosphere'*, which provides the nuclei for rain droplets to form in storm clouds containing millions of tons of water, resulting in torrential rainfall, as demonstrated by the abnormally heavy rainfall which occurred during August 2021, leaving reservoirs full to the brim for the first time in 100 years.

2) **Useless activities with harmful effects on the precious bio' spheric level of breathable air,** the realm of living things, to be banned forthwith: e.g., the deleterious effects of *'Formula 1: Gambling Races'; 'Large scale deforestation'; 'Burning fossil fuel'*, etc.

3) **Imperative: Space-travel** *'Rocket vortical onslaughts'* should be stopped, or limited to the bare minimum to service the existing Space Station Laboratory, until a thorough investigation has been carried out, to ascertain how much damage has been inflicted upon our world's vital protective shields, *'Magnetosphere & Ozonosphere'*, by the immense power of scouring explosive energy exerted in continuous experiments, with rocket missiles accelerating to a velocity of 37,000 miles per hour, through Earth's fragile atmospheric shells.

Ray of hope: Human exploitive and aggressive behaviour have left little room for **Uriel, *'God's Angel of Peace'*,** to make his presence known, during centuries riven by war. The truly remarkable phenomenon of the 21st Century must be, that minds have been opened to see oneself and the world in a more caring way. Theologians, Educators and Scientists, appear to be converging in a more analytical, and ecologically inspired manner of thought, acknowledged as *'The Information Age'*, when all can agree that it makes good sense for all living beings to look after each other: *'For none are safe until all are safe!'* At last, all may be speaking and thinking as one, inspired by *'Uriel's peaceful presence'*. **Maxwell's intuitive hunch, that *'energized fields of flow' appear to be imbued with a regulatory inverse power of order and control,* may now assume a more profound sense of meaning, in this new *'Age of Enlightenment'*, pointing**

the way to make wise decisions in managing impending problems, such as 'Climate warming'.

With calm heads, and 3 strands of essential knowledge, it should be within our grasp to make wise decisions, to mitigate the anticipated detrimental effects of 'climate warming':

• *There is no shortage of energy on Earth. Vast stores are held in the oceans, as well as geothermal supplies, ready for the taking, and free to all, including the poorest nations.*

• *These supplies can be tapped easily and converted to electricity from these limitless stores: (***'In theory, ocean thermal energy conversion (OTEC), could provide 4,000 times the world's energy needs in any given year, with neither pollution nor greenhouse gases to show for it. The ocean is a massive and constantly replenished storage medium for solar energy. Most of that heat is stored in the top 100metres of the ocean, while the water 1,000 metres below -fed by the polar regions-remains at a constant 4 to 5 degrees C. Steam powered turbines drive nearly every coal & nuclear plant in the world, but their steam is provided by <u>burning polluting coal or generating long-lived nuclear waste.</u> OTEC, by contrast, provides steam in a clean and theoretically limitless way.'***): NS: 1 March 2014: 2958: pp.49-51:Helen Knight: 'Sea Change: Oceans of Power'.*

• In addition to the unlimited stores of thermal energy held in the oceans, the kinetic energy in wave momentum and tidal surges has been recognized for several decades, as among the cheapest potential sources of energy. In 1898, Nicola Tesla demonstrated that a boat could be driven by electric power and showed how electric power could be transmitted freely. Since then, it has become feasible to construct island pontoons or ships specially designed for that purpose. New fuels are also being developed, such as a derivative of Ammonia, which promises to be an important clean fuel of the future. Made of Nitrogen, which constitutes 78% of the atmosphere, plus Hydrogen, it burns leaving only clean water as a residue. *(NS: 3 Aug. 2013: 2928: p.22: Hal Hodson: 'Out of Thin Air: Crucial to Our Civilization'.*

Following technological developments of solar panels, capable of transforming solar radiation efficiently into electric power, nuclear plants have become outdated. Deserts begin to appear as lands of promise, instead of wastelands. With vast space for batteries of solar panels, they could be transformed into productive, fertile manufacturing hubs of the future, supplied with cheap electrical energy, producing desalinated clean water, and equipped with prefabricated housing units- designed for tropical regimes, using chemically formulated new materials.

Warning: If the signs of increasing seismic and volcanic activity, we are witnessing (e.g., Las Palma: Canary Is.: Sept.2021: & Iceland: Fagradalsfjall Mt.: volcanic eruption & earthquake crisis: Feb.-Sept. 2021), should prove to herald an episode of basaltic outflow, which could last for thousands of years, the last thing we would want would be the detonation of stores of nuclear warheads in underground bunkers, engulfed by molten magma at a temperature of 800° C. The same caution applies to *Nuclear Plants.*

The Ultimate Folly: Amateur experimentation with '*Nuclear Fusion Energy,*' involving the control of '*Plasma*', a phase-state unknown on Earth, should be banned at World Government level, recognized as a potential lethal hazard.

Diagram: Variation of fluid velocity from the centre

Uniform rotational field of flow : Vortical field of flow

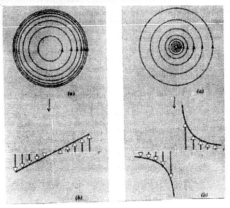

Note:- The curvature of the Earth is constantly changing, flexing and bending, never fixed.

Pythagorean Theory:
Radius=3956 miles
a=distance
b=radius
c=drop radius

Idealized concept of Planet Earth

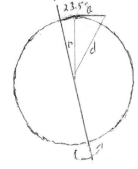

Reality of lumpy 'Geoid Earth'

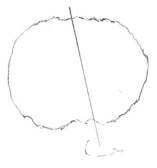

24

The key to understanding the effects of vortical energy flow, regulated by variations in sinuosity and velocity of flow in *synoptic weather patterns': Mid-Latitude Climatic Regimes.*

N. Hemisphere   S. Hemisphere

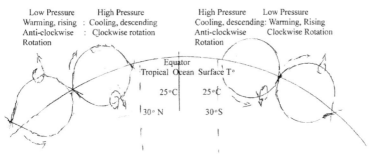

Low Pressure	High Pressure	High Pressure	Low Pressure
Warming, rising : Cooling, descending	Cooling, descending: Warming, Rising
Anti-clockwise : Clockwise rotation	Anti-clockwise Clockwise Rotation
Rotation	Rotation

Equator
Tropical Ocean Surface T°

25°C   25°C

30° N   30°S

Phenomenon of the 21st Century: Realities of Existence

We have been bombarded with adverts telling us that the world is a better place, in terms of living conditions, because of advances in scientific discovery and associated technological developments, based on the gadgets we have come to rely upon for communication, such as the telephone, television, cinematography, etc. But our perception of the world about us and its connections tells us otherwise. We know and learn from the accounts of others' experiences, that for most people the world does not appear to be improving as a better and safer place. Contrary to the claims of those who reject the concept of a supernatural creative *'Mind of God'*, in favour of the belief that the *'human brain/ mind'* alone possesses the rational ability to solve any problems of organization and management, associated with any environmental hazard we may encounter, we know instinctively as well as experientially, that this is far from the truth.

In fact, dismissal of spirituality of mind has the debilitating effect of dimming the cognitive faculty of understanding patterns of connection and regulation, in the chain of communication by which Knowledge of discoveries and experience is passed on and transferred, as a *'process of cognitive development': (Theory of Learning).* Even animals appear to retain an instinctive impulse to pass on valued knowledge, gained from past geographical experience, as an *aide memoire* to their offspring, to cope with any exigencies which may arise in the course of life. For example, just watch how the Dolphin, noted for its sophisticated social sense and behaviour, teaches its offspring to obtain food efficiently, whilst avoiding environmental hazards such as becoming marooned in shallow water.

('We learn from stories- the context of the phenomenon of transferred experience. Context is linked to another undefined notion called 'meaning'. Without context, words and actions have no meaning at all. This is true not only of human communication in words, but also of all communication whatever, of all mental process, of all mind, including that which teaches the sea anemone how to grow, and the amoeba what he should do next'): Bateson, G: **(1)**.

Science never proves anything: Scientific theorizing becomes particularly enfeebled by the denial or belittlement of the spiritual quality of discernment, imbued in the human mind, as a principle of ordered regulation.

('Most scientists discount the concept of duality, (spirituality of mind and soul in man but it kind of makes sense to take care of the environment. My guess is that one day we'll understand consciousness as just another state of matter. I'd expect there to be many types of consciousness, just as there are many types of liquid, but in both cases, they share certain characteristic traits we can aim to understand': Mark Tegmark: Mathematician and Physicist). **(2)**

This appears to be a materialistic view of life and the world, and not conducive to spiritual qualities of empathy and ecology, as in-built systems of guidance and regulation. Thus, the very systems, needed to maintain stability and well-being in the complex dynamic network of connections, defining **Life processes,** become enfeebled in the same way, the aesthetic quality of mind, which enables us to value Life as precious, and to contemplate, **'The wonder of our being in an exquisitely ordered and beautiful Earth world',** may be diminished. Also, a materialistic view of Life is very different from the recorded perceptions of leading scientific minds, such as Plato and Darwin, who evidently understood that their observations of *Life/World* geographical interactions revealed an innate principle of orchestrated organization. This opening of mind has developed into **Chaos Theory,** and *'The Theory of Universality',* developed by Mitchell Feigenbaum and colleagues at the Santa Cruz Laboratory, which stipulates, *'that there is a principle of order and regulation common to all systems',* heralding the inauguration of a new science, *'Cybernetics: The human mind as an integrated system of control'.*

Charles Darwin avoided the question of the origin of Life, as shown in a letter he is reported to have written to a friend: *'Any attempt to explain the origin of Life is rubbish: you might just as well try to explain the origin of matter!'* In another letter, written in 1882, he affirms: *'I believe that the 'Principle of Life' will hereafter be shown to be part, or consequence of some general law'.* We should also remember that Darwin never subscribed to an atheistic view of *'Life and the Universe'.* On the contrary, it is evident that he recognized patterns of intricately ordered relationship in all his observations of natural events, which as he believed, could only be explained in terms of intelligent information, emanating from a rational mind.

('Darwin, in his genius and farsightedness was right, and that such a theory can now be formulated-- I will attempt to show that Chemistry, the science that bridges Physics and Biology, can provide answers, still in part incomplete, to these fascinating questions. Achieving a better understanding of what life is, may not only tell us who and what we are, but will hopefully provide greater insight into the very nature of the Cosmos and its most basic laws'): Professor Addy Pross, distinguished for his studies in Organic Chemistry, and an authority in the interactions of molecular systems. **(3)**

In the context of transfer of geographical experience, Professor Pross emphasizes that Life systems are shown, by the very molecular reactions which brought Life into existence, and which continue to ensure heredity of self-replication to an astonishing degree of fidelity, to be imbued with meaning and purpose, defined as *'an organizing principle of teleonomy'*. In my perception, as a Geographer, preoccupied with the connections between living systems and what must be acknowledged as our world's remarkably ordered beneficial environment, this confirms that both Life and our Geophysical world must be understood, as interlocking teleologic entities. Therefore, the human mind should be considered as a vital subsystem of conceptual reasoning, endowed with the ability to communicate and interpret perceptions by means of language decryption, *(Maths, Linguistic Codes, Music & other sensory perceptions)*, that make the *'state of Life'* special, and very different from inanimate matter. Thus, **Life must have a purpose, and we are meant to be aware of this.** *(Gk. teleos=complete, from telos= end, purpose. Gk. Kruptos = hidden).*

Organized Complexity: In a small bacterial cell ($1,000^{th}$ of 1 millimetre), are packed, *('a vast assembly of molecular components, $1,000,000^{th}$ of a millimetre, performing a highly integrated network of chemical reactions – in a number of chemical forms- required for the biosynthesis of essential chemical messengers- working in chains of communication and command, to ensure the accurate transmission of instructions for the control and regulation of the cellular machinery, needed to progress by way of energy gradients, designed to maintain the structural integrity and organization of the cell, through the continual utilization of energy- as necessary to empower the functional metabolic chains in Life Systems').* **(4).**

According to the 2nd Law of Thermodynamics, energy in a system tends to dissipate and run down to a lower level (energy gradient), giving way to a more disorganized state, defined as *'increasing entropy'*. But Living Systems in our

Earth-world do not appear to comply with such a so-called law. On the contrary, they appear to be regulated to maintain their highly organized functions, as well as to replicate them when necessary, with an astonishing degree of accuracy, defined as *'a state of homeostasis'. (Gk. en=inside +trope=transformation= change of state: Gk.organon=instrument, engine from IE base=work: homeo=like+stasis=stationary,still=stable).*

Life's Sense of Purpose: All living organisms appear to have been imbued with an awareness of what they are meant to do, as if they have an innate understanding of their existence, as unfolding patterns of development, defined by biologists as *'teleonomy (end state).* Professor Pross argues that, *('In some fundamental weird sense of reality, we are living simultaneously in 2 worlds, each governed by its own set of rules- the Laws of Physics & and the Laws of chemistry, within the inanimate world & the 'Teleonomic Principle' which dominates the biological world---All our interactions with the inanimate world* (Systems of Human Organization), *are based on the recognition that there are 'Laws of Nature' described by the physical sciences, which govern the manner in which the Universe functions').*(5)

However, geophysical systems have their own complexity, ranging from small scale patterns, such as land and sea breeze, *'anabatic and katabatic winds',* as local changes in atmospheric circulation in response to thermal gradients between land and sea; and larger scale patterns, such as the onset of turbulence in the *'Jet Stream' (N. hemisphere)',* cascading into chaotic patterns of *'Rossby Waves'* -leading to violent changes in weather, as tongues of very cold air are drawn to the south, or unusually warm air is driven northwards. The same principle of *'scaling',* by which small non-linear patterns of adjustment- acting as recursive loops of feedback and regulation, connect with larger linear patterns of organization, appears to apply in both cases.

('In the living world complexity is not arbitrary, but specific: Life, as we know it today, is based on a large number of compounds, that are not formed under thermodynamic control. It is enough to think of the most important functional biopolymers- the proteins and nucleic acids. Those are chains with a specific sequence—e.g. The digestive enzyme 'a-chymotrypsin, we have in our gut, is a long chain with 241 amino-acid residues, and the enzyme lysozyme has 129 amino-acid residues. In order to be active, these enzymes must possess their specific sequence ("primary sequence")--that is the very precise order of amino acid residues along the chain- and that applies to each of the very many proteins of our cells. Each step requires 'knowledge' of precisely what occurred in the previous step, in order to ascertain what must be achieved in the next step, defined as 'contingency, all along the chain':
'The Systems View of the World': Capra, F. & Luisi, P. L. **(6)**

If we perceive our own mind in the light of our geographical experiences, as well as in our observations of the working of other minds, to be a natural system of protection and control- like the immune system which exists to protect our body, in order to steer safely through any pathogenic or environmental hazards which may be encountered along the path of Life, it would be unwise to ignore what it is telling us, as a valuable system of guidance. In this respect, the collective effect of scientific and philosophical ruminations through the ages, cannot be considered to have led the way. For example, leading Medics are still inclined to define *'Cognitive Faculties'*, as the working of the *'switch-board of the Brain';* and scientific misinformation can become so myopic, that it becomes nonsensical: *e.g. 'The God Delusion: Life as the creation of "A blind watchmaker".* Who could be so blind, as to undervalue *'the wonder of being alive',* in that way? That is how the significance of valid knowledge, gathered through the ages, concerning Life processes and environmental vicissitudes, together with philosophical and theological speculations about the mysterious spiritual and cognitive faculties of *mind and soul, defined as 'Consciousness'*, may be obscured, misinterpreted, and disseminated as false information. Amongst the philosophers of old, there was much debate about **consciousness,** in terms of the qualities of immanence and transcendence in the human mind: whether the perception of reality only exists in the human mind as an inherent system of communication with the creative Mind of God, or whether the mystery of Life and the attributes of the surrounding Universe were beyond the reach of human understanding and experience.

The easy way out of this dilemma was to dismiss the concept of *'spiritual endowment'*, reducing it to a simple mechanism, which could be explained as the switching of synapses in the brain, with the effect of diminishing or shutting down the capacity of human comprehension. In a similar way, the significance of scientific discoveries may be blocked or impeded in the formation of a chain of developmental knowledge. *(' Knowledge is like a sphere: the greater its volume, the greater its contact with the unknown': Pascal: Pioneer of 'Probability Theory).* **(7).**

The way we see ourselves, and come to assess the environmental quality the world, or region around us, significantly affects our perception of all the relationships and connections involved. As a result, the knowledge gathered from the lessons of geographical experience colours the **cognitive processes** of perception and discernment of the world, and our location, as auspicious or worsening. Judging by the mood of pessimism, which appears to be clouding the mental outlook of our younger generation, as evidenced by increasing addiction to drugs, and the rate of suicides, the potential value of the human mind as a system of guidance and control, must be regarded as a vital part of our biological makeup. Our world home would surely be better ordered, if full use were made of the knowledge we now possess, concerning the protective power available in *'Life's self-replicating immune systems'.* *('Of all the things that a human has to develop, the most remarkable and the most difficult to understand, is the human mind. We still know very little about how the higher functions of the mind work, though it is clear from the specific and predictable effects of different types of brain damage, that the mental phenomenon of mind emerges from the neural activity in the physical brain---When finally complete, the central nervous system consists of tens of billions of individual cells, each of which can be connected to well over a thousand others'):* 'Life Unfolding': James A. Davies: Professor of Experimental Anatomy: Edinburgh. **(8).**

Integrated Information Theory: The Human Mind as a Central System of Coordination: During the 1960's, developing knowledge concerning geographical connections as interactions between Life forms (plant, animal and human), and environmental conditions, crystallized into an understanding of the human mind as, *'the process of Life',* by which our mental faculties, (learning, memory, decision - making, etc.), enable us to perceive the key criteria of Life (pattern, structure & process), as innate patterns and relationships of organization. **(9).**

Neurobiologists became aware that neural responses *'could no longer be interpreted as a stage by stage process of information in the brain',* which led to a shift in focus , during the 1980's, to the emergent properties of neural networks. This became known as 'The Santiago Theory' of cognition. *('Cognition entails a continual bringing forth of a world through the process of living---the interactions of a living system with its environment are cognitive interactions, and the process of living itself is a process of cognition--- Cognition involves the entire process of Life, including perception, emotion & behaviour. Cognition is an integral* *part of the way a living organism interacts with its environment. It does not react to environmental stimuli through a linear chain of cause and effect but responds with structural changes in its non-linear organizationally closed autopoietic network. In human beings, the bringing forth of such an inner world is intimately linked <u>to language, thought and consciousness</u>'):* 'Santiago Theory of Cognition': Francesco Varela. **(10).**

Viewed in this way, the human mind must be regarded as a control system to be cherished and heeded with the utmost attention for our own good, for it points the way of promise to follow. In this context, we are learning more about the living body's immune system, as an amazingly powerful protective shield, especially equipped with detailed chemical/biological knowledge, for the purpose of warding off pathogenic attacks, including bacterial and viral *pandemic disease.* Even our mucous glands seem to know how to provide healing antiseptic saliva when needed. Yet, we still know so little about ourselves, and the mystery of finding ourselves aware of being *alive.*

Every cell in our body knows how to make us, from a DNA language code of 3 billion units of information? Any slip or change in that precise sequence of information could lead to thousands of genetic disorders, but the spirit of life within you is so ontologically directed to protect you, that the rate of error in all the billions of chemical-biological interactions involved, is known to be less than one in a billion. How is it that my body seems to know more about me than I know myself? And what is more bewildering, my **Mind,** which cannot be perceived merely as the operation of my brain, is capable of keeping me aware of all this complex network of existence, as well as providing information about all that is going on in the world.

('It is no longer a secret that inherited notions of matter and the material world have not been able to sustain the revolutionary developments of 20th century Physics & Biology---complex systems such as living organisms, societies and human beings, could according to this reductionist view, be explained in terms of chemical components & their chemical reactions – 'matter' gradually lost its use in science to be replaced by more robust & measurable concepts of mass (inertial, gravitational, etc.)--3 new developments forced the downfall of the 'Matter Myth'- 1) Einstein: Special Relativity (1905) & General Relativity (1950)-principle of equivalence of mass & energy: 2) Quantum Theory, describing a fundamental level of reality, replacing the assumption that the so-called laws of nature are the most basic level of description, but rather 'Information' as the foundation on which physical reality is constructed & 3) The human brain/mind as the most powerful information-processing system known—the relationship between mind & brain is the oldest problem of philosophy, 'The Information-Matter dichotomy'-- Conscious awareness remains a stubborn mystery- Is the Mind working as a Quantum (Qualitative Processor?: 'Information & The Nature of Reality': Ch. 1:'Does Information matter?': 'Information & the Nature of Reality': 'Introduction: Does information matter?':Paul Davies & Niel Henrik Gregerson. **(11).**

The Enigma: Understanding the mystery of Life in the Universe: The view of the Universe, as a 'gigantic heat engine', generating energy to run the Universe, as defined by Maxwell (*'Wave Functions'),* and Boltzmann *('Probabilities'),* has given way to a more enlightened view of the meaning and significance of *'Information',* emerging in the *'conscientious awareness'* of the Human Mind, concerning the mystery of the existence of living beings. Thus, what was considered as accurate information, derived from observations of events happening in the world around us less than a century ago, has blossomed as *'purposeful structured information'*, with the effect of changing perception of a material World/Universe, in terms of *'mass, energy & motion'*, to a more profound conceptual understanding of, **'the reality of existence in a Universe functioning and emerging, according to a " Supreme Principle of Organized Order".**

('Beyond the Quantum World Lies the Realm of Information. Information is inextricably linked to some physical system—it is carried in the arrangement of molecules in a strand of DNA, enabling the propagation and evolution of Life. It is encoded in the charge of a capacitor in an electric circuit – allowing us to build the information storage and processing facilities we call computers. It is written into the quantum state of a photon of light, allowing conversations to be sent through optical fibres. Wherever information exists, it takes a physical form. This has become known as 'Landauer's Principle': *Rolf Landauer:IBM 1991): '20 Big Questions in Physics: pp. 72-82: Michael Brooks.* **(12)** Information is processed for transmission, in communication and dialogue, by *Language,* which is a fundamental process of *Cognition,* the acquisition of empirical factual knowledge. Although generally taken for granted, this faculty of *Mind* should be regarded and valued as a *Systems Phenomenon,* linking Cognitive and Life Processes. **(13)**

In this respect, mainstream scientific thinking seems to have overlooked *'Theories of Education & Learning'*, which are rooted in the capacity of the Human Mind, to process conceptual thoughts by means of, **'a modular organization for language decryption'. Language is the platform of understanding therefore, by which the mind is opened to comprehend what may be perceived as the *'connections and realities of existence'*. Life and the endowment of Language capacity came into existence and go hand in hand together. (14)** From the systems point of view, the understanding of Life

36

and the associated connections and relationships of geographical experience, stems from the knowledge gained from the recognition and study of patterns in the events which unfold and shape perception of the world and other people. In this context, the meaning and significance of knowledge can become distorted, or even twisted to become misleading information. For example, concepts such as *'Democracy'*, originally established by Pericles the Great, arguably one of the most successful of men in whatever role he took the lead when in public office, with the express intention of extending the right to take part in government decisions to the working class. Hitherto, it had been jealously regarded as their prerequisite by the wealthy ruling elite, to make decisions concerning systems of human organization and government. He had to stand firm against a barrage of hateful opposition. Have things changed? As yet, no country can claim to have achieved a truly *'Democratic Ethos'*, of government. On the contrary, the very meaning of *'Democracy'*, is frequently subverted to mean freedom of speech and action to do as you like, regardless of others. It is salutary, to be reminded how important it is for language to convey meaning, **in the light of the spiritual qualities of Mind, 'Truth and Love'.** Ecological principles do not go hand in hand with the baser motives of human nature, **the pursuit of power and wealth,** which have become dominant in our world, dragging us to the precipice of destruction.

Phenomenon of the 21st Century: Opening of Mind to comprehend 'The Realities of Existence: This century will be remembered for the astounding scientific discoveries of' *'The Human Genomic code', the complete set of genes or genetic material carried as knowledge in each cell of the body,* and of *'Quantum Electrodynamics (Quantum Theory)', concerning the Mathematical Laws which appear to underpin the Universe.*

This astonishing network of knowledge revealed that the very molecular systems defining biochemical reactions in the metabolism of living things, as well as the interactions observed in geophysical environmental systems, **must have been intricately organized to make Life on Earth possible.** *('All things are linked, including properties of atoms, and the bonds holding them together. We have reached the stage of mapping the Universe, when the grand outlines are coming into focus—our emergence and survival depend on very special tuning of the COSMOS: If E were 0.006 or 0.008 we could not exist')* : Note:- *E –This value is 0.007, and it defines how firmly atomic nuclei bond together and how all atoms on Earth are made. It controls the power from the sun, and how hydrogen is changed into all the atoms of The Periodic Table in the stars)*:*'Just 6 Numbers': Ch4: 'Stars , The Periodic Table, and E': pp.45-57:Lord Martin Rees: Astronomer Royal:* **Fig. 3: 'Molecular Forces to make Life on Earth possible'. (15).**

These mind-boggling revelations have rocked the sciences to their foundations by affirming that **nothing in the Universe can now be considered as fixed and substantially material in nature.** Also, the concept of fixed immutable *'Laws of Physics'* can no longer be upheld. For example, the *'Newtonian 2nd Law of Thermodynamics'*, postulates that *'as energy in a system runs down and is dissipated, so a state of increasing entropy (disorder) is generated, 'sliding inevitably to a "Cosmic Heat death".* We now know that **Life on Earth appears to be like a battery under charge, capable of storing energy and replicating itself.**

In this connection, it is relevant to take note, that Newton may not have been the totally empirical scientist he was thought to be. He is now known to have been a committed member of the secret *'Order of Alchemists',* in search of ultimate wealth and power. A letter, written in coded Latin to a fellow alchemist has been discovered, in which he warns, *'This knowledge must be kept secret, because it could destroy the world'.* It is possible that he may have stumbled upon the immense power of *'Nuclear Fusion',* during his study of *'Energy and the Photon of Light'.* In this way, valid knowledge may be withheld or obscured. However, Newton's *'3ʳᵈ Law of Motion',* in respect of the workings of the physical world, ironically predicts that there is always an opposite and equal reaction. Pernicious influences can also be regarded as a form of energy, which act to condition the way people come to envisage their future world, and affect the outlook towards other people, either with a reaction of hate or trust. Thus, baleful forces may be countered by an equal and opposite reaction, acting as a protective shield, in the spiritual dimension of the human mind.

Cognitive Immunological Reaction: In this context, the molecular structure of DNA, now understood as the building blocks of Life and defined as the Human Genome, first discovered by Friedrich Miescher (1868), investigated by Rosalind Franklin (1950), and published by James Watson and Francis Crick (1953), opened the understanding of the Human Mind further, as a *'System of Control and Guidance',* capable of steering the way amid *'The Realities of Geographical Existence,* as well as exerting a healing influence. *('In the perspective of human history, the main object of scientific research on the brain (mind), is not merely to understand and cure various medical conditions -important though this may be, but to grasp the true nature of the human soul'):* 'The Astonishing Hypothesis': Francis Crick. **(16).**

This reflection, concerning the *'the principle of Life in humans and animals',* signifying the experience of *'animate existence* (knowing and feeling the state of being alive), may be taken to imply that *'soul'* could be understood as an entity distinct from the body: but the words of Christ tell us that *'body and soul are bound together in a spiritual state of existence',* from the very inception of Life, as ordained according to the *'Will and precepts of the Mind of God'.*

('Anyone who loses his life for my sake, that man will save it. What gain then is it for a man to have won the whole world and to have lost or ruined his very soul (self)?'): Luke 9.23-36. (17)

In this respect, it is evident that scientific thinking about the mysterious nature of Life, has been compelled to have second thoughts, having diverged from Theological doctrinal teaching about *'The spirituality of existence' for several centuries*

('When we find a unifying theory, linking the laws of Physics, especially the fundamental probing theories of Quantum Mechanics, we shall see the Mind of God': and *'The unit of Information is more fundamental than anything else- even quantum particles'): David Deutsch: Pioneer of Quantum Mechanics: ' Why we need to reconstruct the Universe': NS. 24 May 2014: 2970.* (18)

Professor Deutsch (Dept. of Atomic & Laser Physics, Clarendon Laboratory, Oxford Un.), was quite right, in spotting that prevailing concepts concerning the *'Organization and Ergonomics' of the Universe, and the state of Life in the World as part of it,* has led the sciences to a quandary of confusion.

The revelation of *'Information'*, as knowledge contained in the emergence of, *'atoms (probabilities),* in the form of *'molecular systems',* required to direct and regulate chains of contingent chemical/biological reactions, has opened a profound field of understanding in the Human Mind. Like St Stephen, who was graced with a vision of Heaven, we have received a glimpse of the *'Mind of our Creator'.* We now know that all things we perceive are in a state of fluid dynamic change, including the very cells of our brains as we read, write, and communicate.

Qualitative Electro-dynamics: This spiritual advance of understanding continues to unfold, acting like a *'Quantum Processor in the cognitive faculties',* developing as the new science of '**Cybernetics**', focused on the **Human Mind, as an**

integrative system of control. *('From the time of the ancient Greeks and before, it had been believed that there must be one set of laws for the Heavens, and a completely separate set holding here on Earth—with Quantum Theory, we seem to have reverted to a scheme like that conceived by the philosopher-scientists of ancient Greece'):* 'Shadows of the Mind': Sir Roger Penrose. **(19).**

This observation is fundamentally significant. It confirms that educated minds of the past, especially, the Jewish Scribes writing the Scriptures in 850 BC., appear to have been well aware that what is perceived as empirical scientific knowledge, could have no meaning without the cognitive capacity of the Human Mind to *know itself in dialogue with the Supernatural Mind of our Creator God',* call it *'resonance of energy'; 'energy gradient'; 'entanglement'; 'Semiotic and Semantic Knowledge',* or what you will. We are clearly being compelled to take stock of our values, and the resulting state of our home-world.

Dialogue in communication with the Mind of God: It appears that there are 3 threads of understanding in the communication between our Human Mind and our Creator:- **1)** Epistemological *(valid knowledge): 2) Ontological Understanding (the meaning of what is known): & 3) Ontogenic purpose (inspiration of wise decisions & practice of "Deep Ecology".*

Similarly, *'3 Realities of Geographical existence and experience',* may be distinguished:- *1) Perception of the Human Mind, providing that it is based on valid knowledge: 2) Life Processes, including Cognitive Processes, which cannot be separated; & 3) Environmental Parameters (influences-good or bad).* **Realities of Geographical Experience, seen as the unfolding effects of geographical events**: In these interactions, it is essential to take account of the complexity arising from the vast discrepancy between the timescales of 'Geophysical fluctuations', and the events unfolding in 'Geographical experience'. Knowledge harvested from the records of human geographical experience is subject to the limitations imposed by the short timescale of human life, and the 'biological clock'. The preparation of Planet Earth for the existence of Life, as recorded in the geological evidence in the rocks, covers a time span of over 500 million years.

Fig. 1: **'Connections between Geophysical & Life Systems:** Comparison of Time-lines relating to 'Oxygenation of our Earth-world', with the span of knowledge concerning the complex network of dynamic intricately linked interactions involved, would indicate that *'Time' and 'Space', can be regarded as illusory concepts.* Human memory fades rapidly, and becomes garbled and confused in the span of a few millennia, whereas the geological records tell us that oxygenation of the atmosphere to make Life possible on Earth, took millions of years, until the first life forms appeared 350 million years ago. Even then, the geological evidence tells us that it took hundreds of million more years, until the interactions of the biomass with atmospheric/oceanic systems had achieved a dynamically stable state of *'homeostasis',* capable of acting as a Life support system.

'**Fig.2: 'Carbon Cycle & Photosynthesis'**: *('We need to be able to understand Life's complexity, and the global characteristics associated with that complexity—we are far from being able to do that'):* 'What is Life?': Ch.6: 'Crisis of Identity': pp. 111-121: Addy Pross. It is really astonishing, as Einstein remarked, that the Human Mind is capable of comprehending what we cannot fully understand. I believe that it was for the purpose of making mankind fully aware of the damage inflicted upon our Earth world's Life support systems, by ruinous conflicts and reckless exploitation, that the Human Mind is being compelled to take stock of the harm done to our home-world.

Fig.3: 'Molecular forces must have been fine-tuned to make Life possible', demonstrates without any shadow of doubt that Life and the Earth-world came into existence as a wondrous *'ordered system of Creation'.* Note that this is a remarkable **comprehensive view of 'Life on Earth from a "systems view of the Universe", confirming in the language code of Maths, the unity of all phenomena in existence, according to a universal "principle of order".** *('The Universe is written in the Language of Maths':* Galileo de Galilei (1564-1642): Polymath and "Father of Science"): Also, (*'All things created are brought into being, directed unto their appointed end':* 'Summa Theologica, Quinque Viae': St. Thomas Aquinas (1225-1274).

Fig.4:'Realities of Geographical Existence': Note:- The phenomenal cognitive capacity of the **Human Mind,** capable of functioning as a *"Quantum Processor",* aware of being a part of **a complex network of interactions,** which cannot be fully understood in its entirety.

(' The reason I talk to you in parables is- you still listen and listen again, but not understand, see and see again, but not perceive—be healed by me!': Jesus Christ): Matt. 13. 10-17.

(Gk.synapse from hapsis=joining, from haptein=connect:

dialogos from dialegesthai=discourse, conversation:

entropy from en= to be, the state of being +trope from tropos= turn from trepein =change, transform= ending of all things and ultimate disorder:

empeirikos from empeiira= trial, experiment, experience, observation: ergon= work, the way things work:

ontogenic from on=being, ontology= science or study of being, existing)

Figure 1: Connections between Geophysical and Life Systems:
Time-lines in advances of science:
a) Geological record of Oxygenation of the Earth-world in preparation to sustain Life:(*After Dr. N. Lane: 'Oxygenation:The molecule that made the world': OUP 2016*).
b) Major scientific discoveries during the last 2,000 years:(*AfterT. Jackson:100 breakthroughs in Physics: Shelton Harbor Press: New York*)

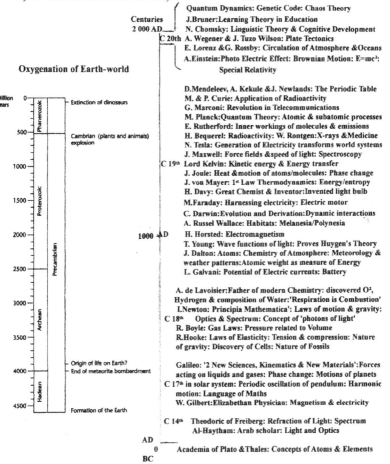

Discoveries.

Centuries
2 000 AD
- Quantum Dynamics: Genetic Code: Chaos Theory
- J.Bruner:Learning Theory in Education
- N. Chomsky: Linguistic Theory & Cognitive Development
- C 20th A. Wegener & J. Tuzo Wilson: Plate Tectonics
- E. Lorenz &G. Rossby: Circulation of Atmosphere &Oceans
- A.Einstein:Photo Electric Effect: Brownian Motion: $E=mc^2$: Special Relativity

Oxygenation of Earth-world

- D.Mendeleev, A. Kekule &J. Newlands: The Periodic Table
- M. & P. Curie: Application of Radioactivity
- G. Marconi: Revolution in Telecommunications
- M. Planck:Quantum Theory: Atomic & subatomic processes
- E. Rutherford: Inner workings of molecules & emissions
- H. Bequerel: Radioactivity: W. Rontgen:X-rays &Medicine
- N. Tesla: Generation of Electricity transforms world systems
- J. Maxwell: Force fields &speed of light: Spectroscopy
- C 19th Lord Kelvin: Kinetic energy & Energy transfer
- J. Joule: Heat &motion of atoms/molecules: Phase change
- J. von Mayer: 1st Law Thermodynamics: Energy/entropy
- H. Davy: Great Chemist & Inventor:Invented light bulb
- M.Faraday: Harnessing electricity: Electric motor
- C. Darwin:Evolution and Derivation:Dynamic interactions
- A. Russel Wallace: Habitats: Melanesia/Polynesia
- H. Horsted: Electromagnetism
- T. Young: Wave functions of light: Proves Huygen's Theory
- J. Dalton: Atoms: Chemistry of Atmosphere: Meteorology & weather patterns:Atomic weight as measure of Energy
- L. Galvani: Potential of Electric currents: Battery

- A. de Lavoisier:Father of modern Chemistry: discovered O^2, Hydrogen & composition of Water:'Respiration is Combustion'
- I.Newton: Principia Mathematica': Laws of motion & gravity:
- C 18th Optics & Spectrum: Concept of 'photons of light'
- R. Boyle: Gas Laws: Pressure related to Volume
- R.Hooke: Laws of Elasticity: Tension & compression: Nature of gravity: Discovery of Cells: Nature of Fossils

- Galileo: '2 New Sciences, Kinematics & New Materials':Forces acting on liquids and gases: Phase change: Motions of planets
- C 17th in solar system: Periodic oscillation of pendulum: Harmonic motion: Language of Maths
- W. Gilbert:Elizabethan Physician: Magnetism & electricity

- C 14th Theodoric of Freiberg: Refraction of Light: Spectrum
- Al-Haytham: Arab scholar: Light and Optics

AD
0 Academia of Plato &Thales: Concepts of Atoms & Elements
BC

Million years axis (left): 0, 500, 1000, 1500, 2000, 2500, 3000, 3500, 4000, 4500
Phanerozoic — Extinction of dinosaurs; Cambrian (plants and animals) explosion
Proterozoic
Precambrian
Archean
Hadean — Origin of life on Earth?; End of meteorite bombardment; Formation of the Earth
1000 AD

Note:-Age of Earth: 4.6 billion years
Oxygenation to support Life: 3.8 b. y.
3 Essential Biomolecules for Life to exist:- O^2, H^2O,& Chlorophyll (6 complex molecules of C,H,O,N&Mg) : 5 Groups imbued with connectivity and purpose? Which came first?
Did Life commence with ordered complex connectivity?

Figure 2 : The Carbon Cycle and Photosynthesis:The Earth-world's Life Support Systems

Diagram: Preparation of the Biosphere by processes of photosynthesis to support respiration in Life systems: *(After Richard Cogdell: N.S.: 2 Feb. 2013:2902; 'Instant Expert 30 : Photosynthesis': & Data from New Science Publications).*

Note:-

The amount of CO^2 in the atmosphere at present is relatively small: ·4% by volume;·062 by weight, 401 pp million. This is about the same as in the Late Carboniferous Period , when CO^2 content declined to about 350 pp million, from a peak in the Early Carboniferous of about 1,500 pp million. Both the volume of CO^2 and O^2 reached their highest peaks in the Early Carboniferous, resulting in an average world temperature of about 20°C (68°F), which greatly energised photosynthesis and enriched the Biosphere with oxygen, leading to a luxuriant growth of vegetation- turning an atmosphere toxic to Life into a Life-sustaining environment.

Only the Late Carboniferous and our present Quaternary have experienced CO^2 levels less than 400 pp million. The present average world temperature is about 58° F.

The oceans contain about 50 times more CO^2 than the atmosphere, and about 10 times more than that held in plant and soil carbon stores.

The amount of CO^2, estimated to be contained in the atmosphere, is about 2 trillion tons, a rich source of carbon as a new material of great potential.

The chief cause of global warming is methane (CH4), 23 times more pollutant as a 'greenhouse gas'.

Atmospheric Oxygen levels %

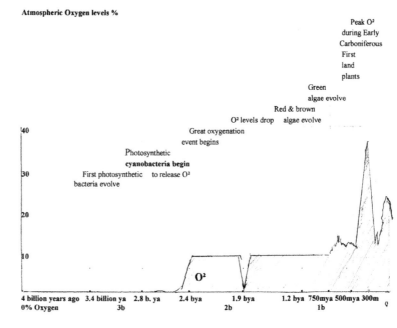

Figure 3 : Molecular Forces must have been 'fine tuned' to make Life on Earth possible:
(Martin Rees:Astronomer Royal:RS. Research Fellow: 'Just 6 Numbers': Cambridge 1999)

Mathematical laws underpin the fabric of our Universe. All things are linked, including the properties of atoms and the forces holding them together. The very existence of atoms depends on the forces and particles deep inside them. Planets, stars, and galaxies are controlled by the force of gravity. Everything takes place in an expanding universe.

The Numbers:

N : One million,million, million,million,million,million : this measures the strength of forces which hold atoms together, and governs how all the atoms on earth are made. If it were smaller, only a small universe could exist, and only for a short time., and creatures could not grow larger than insects.

E : This value is 0.0007, and it defines how firmly atomic nuclei bind together, and how all atoms on Earth were made. It controls the power from the sun, and how hydrogen is changed into all the atoms of the Periodic Table in the stars.

O: Omega measures the amount of material in our universe (gas, galaxies, and dark matter). It tells us about the relative importance of gravity and expansion energy in the Universe. If this ratio were too high the Universe would have collapsed: if too low, no galaxies or stars would exist. So the initial speed of expansion must have been 'finely tuned'.

L: Lamda was found in 1998. It is an unsuspected new force of cosmic antigravity. It controls the expansion of our Universe (on the large scale of over One billion light years. It will get stronger but is small enough now to let stars and galaxies form.

Q: Represents the ratio of 2 fundamental energies, and is about 1/100,000 in value. If it were smaller the Universe would be inert and structureless: if larger, it would be a very violent place where stars and solar systems could not exist, and would be dominated by vast black holes.

D: This is the number of spatial dimensions in our world, 3. Life could not exist if D were 2 or 4. Time is a 4th dimension, but it is different because it has a built in arrow towards the future.

Note : If any of these numbers were different, there would be no Life and no stars. It is astonishing that an expanding universe can be specified by just a few numbers: this means that they are a blue print, a code left for us to decypher---- a communication of Mind.

What do we see by zooming up (each step 10 times further away) : In 10 steps we would see the entire Earth, with continents, oceans and clouds: in 25 steps we would see to the furthest limit of the Hubble telescope- the limit of our Universe.

Zooming in smaller, we begin to see the finer texture of human tissue, and then into a human cell. We can see that there are 100 times more cells in a human body than there are stars in our galaxy. No real telescope can probe within the atom,where a swarm of electrons surrounds the positively charged nucleus: but in 2014, Professor Kaminski and team developed the Simulated Imaging Microscope, which opens the possibility of observing the functions within the cell.

We are each made of 10^28 to 10^{29} atoms, that means that we, as part of Life in our Earth-world, are midway in scale between the masses of the atoms and the atoms of the stars. Also, the number of atoms in a human being is about the same number of human beings to equal the mass of the sun.

The Cosmos (1 billion metres in diameter)- Living Organisms- A Molecule(1 billionth of a metre) .

47

Figure 4: 3 Realities of Existence from a Systems View, linking Life and the Earth-world:-

Mind: Consciousness<Language<Cognition<Communication<Freewill<Spiritual Awareness
Environment<Geophysical<Oceans<Atmosphere<Biosphere<Troposphere<Stratosphere
Life: Quantum Biology<Oxidation<Photosynthesis<Human Organization<Ecological awareness

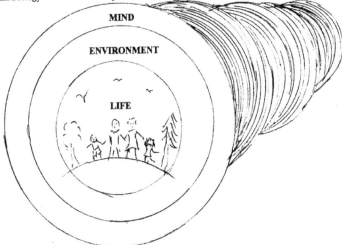

Equations linking systems of energy transfer in a lattice of Geographical connectivity:-

Planck's Constant of Proportionality:$E=hv$: Energy of an electromagnetic wave is proportional to its frequency<fluctuations>aperiodicity+regularity in systems>orderly functions.

Einstein: $E=mc^2$:- *'This equation is the very root of Life. It describes the fundamental nature of Reality, revealing how deep is the familiar notion of matter, as an illusion: all atoms are the same stuff, with properties only definable in terms of functions in systems- probabilities rather than things':(Michael Brooks:'20 Big Questions in Physics:pp.160).* c/f:Erwin Schrodinger &Niels Bohr.

Rossby No.: $Ro=U/fL$: The ratio of the product of the **Coriolis parameter and the inertial centrifugal force** of the Earth's motions, whereby fluctuations and oscillations at **scales of time, space and velocity,** can be taken into account, in comprehending the complex fluid dynamics linking **Atmospheric & Life systems in the Biosphere.** Thus, at the Geological time scale, patterns of vorticity may be seen as related to Eras of Climatic Change.

'Maxwell's Demon': James Clerk Maxwell's hypothesis, *('Theory of Heat';1872)*, showing how the total entropy of the universe could be decreased, in contradiction of Newton's *(2nd Law of Thermodynamics')*, which stipulates the inevitable run-down of energy in the Universe, with *'increasing entropy',* to a state of complete disorder. His thought experiment pondered that, whilst energetic molecules (strong photo-electric effects), can be considered as hotter and destructive (e.g *'Ultra Violet radiaion'),* slow molecular reactions, taking another path, could exert a cooling and ameliorative effect: i.e. acting as an immunological and corrective response. Lord Kelvin called this seemingly circular and self-contradictory puzzle,'Maxwell's Demon': *('Journal of Nature'; 1874).* Note:-1) Earth's Geophysical Systems are able to store energy, like a battery: 2) Life Processes also exhibit an adjustive immunological responce, at times of stress and crisis.

The World in Disarray at a Crossroads Dilemma

Without any shadow of doubt, something remarkable and very strange has swept like *'a new broom'*, throughout the collective human consciousness, during the decade leading to the 21st century, completely changing perception of the human condition in relation to the state of the world. We are reminded that **'Knowledge is like a sphere, the greater its volume the greater the contact (understanding), with the unknown':** *'Pensees': Blaise Pascal: Pioneer of Probability Theory.* **(20).** It is as if the cobwebs of neglect, together with the dust of decay resulting from centuries of reckless environmental degradation and destructive human conflicts, are being swept away. Are we not witnessing a phenomenal cathartic *'spring-cleaning'* of our home-world, impelled by the extraordinary revelations of advances in science, as well as enlightened theosophical insight?

(Gk.katharsis from kathairein =cleanse from katharos =pure).

Advances in scientific knowledge: Plato (429-397 BC), became riveted on the meaning of shapes, such as the cube, tetrahedron, octahedron, dodecahedron & icosahedron, *'The Platonic Solids',* convinced that such structures must have a reason of profound significance. He also was keenly aware of the transient nature of substances, such as the phase states of water, as a gas, liquid and solid. He pondered on the properties and behaviour of *water*, opining that it must be considered as

a *'precursor of Life',* showing an extraordinary cognitive ability to comprehend the *'unity of all things',* long before phenomenal advances in scientific understanding during the 20th Century. All the sciences appeared to be converging, to the point of opening the way to more profound discoveries, concerning *'The Properties of the elements', 'The Genetic Code', 'The Human Genome', The Order of Quantum Electrodynamics', & Quantum Biology of Cellular Functions'*, a complex network of interactions beyond understanding, yet open to comprehension, with the implication that, ***'The Human Mind may be capable of acting as a Quantum Processor of epistemological knowledge'***.

('Everything we call real is made of things that we come to understand cannot be called real: Isolated material particles are abstractions, their properties being definable and observable only through their interaction with other systems'): Niels Bohr, Pioneer Nuclear Physicist.**(21)**. **Bohr** (1885-1962), discovered that electrons travel in separate orbits around the nucleus, and that properties which define an element are determined by the number of electrons, and his research opened the way for further fundamental understanding of *'Atomic Structure'* by **Heisenberg** (1901-1976), known as *'The Founder of Quantum Field Theory'*, whose experimental work concerning the *'kinetic energy'* of emitted light, revealed that electrons are dislodged only when light exceeds a certain frequency. *('The world thus appears as a complicated tissue of events in which connections of different kinds alternate, overlap, or combine and thereby determine the texture of the whole; what we observe is not nature itself, but nature exposed to our method of questioning'):*Werner Heisenberg(1958): Pioneer of Theory of Uncertainty.**(22)**.

James Maxwell Clerk (1831-1879), had established a theory in 1865, uniting all electric and magnetic phenomena, explaining the relationship between electricity & magnetism, and predicting that visible light is caused by electro-magnetic waves, whereas Einstein thought that a light beam is a collection of particles called *'photons'*, rather than a wave through space. **(23)**.

The Photoelectric Effect: In 1887, Rudolf Hertz discovered that when ultraviolet light shines on two metal electrodes, the voltage between them at which sparking takes place changes, whilst working on the transmission of radio waves. In 1905, this effect was explained by Albert Einstein as the transmission of energy by particles, *'quanta'*, at the same time as he published his paper on *'Special Relativity'*. These discoveries opened further the understanding of the properties of atoms, molecular systems and the perception of matter as solids, leaving Physicists in a state of quandary, perplexity and uncertainty. Whereas *'Classical magnetism'* had predicted that continuous light waves transfer energy to electrons, which would then be transmitted when they had accumulated enough power, experimental results showed that only when light exceeds a certain frequency, are electrons displaced, regardless of the intensity of light, or duration of exposure. Now, nothing can be considered as solid. Rather all things begin to look unreal, and not as we used to think of the world.

Yet, just as Plato wondered if the material shapes and structures which define the world as we know it, could be a but a reflection of a more perfect world (*Universe*)- like shadows on the wall of a cave, the stream of knowledge leading to the discovery of the *'Photoelectric Effect'*, continues to gather impetus. As well as giving rise to a new revolutionary advance in technology, with the development of *'Laser Beam'* devices, it has caused Physicists to reconsider whether there could be *'Multiverses'*. In this context, David E. Deutsch, pioneer of *'Quantum Theory: Quantum Field Information'*, has proposed **'A Theory of Everything'**, encompassing the convergence of theosophical and scientific perceptions, and emphasizing that Life and world systems must be viewed **(*'as emergent*, *rather than reductive'*)**: *'The Fabric of Reality': Professor David E. Deutsch*. **(24).**

The essential message of *'Quantum Theory'* is that, **the Phenomenon of Life must be understood as part of a** systemic unity of creation, brought into being and energized by an omnipotent intelligence, infused by a 'Principle of Universal Order', however defined: whether as the *'Word and Will of God', 'The Photoelectric Effect' or 'Energy gradients in molecular knowledge-regulated systems of Order'.*

Theosophical Wisdom: Thus, in 2019 Pope Francis, Ph.D., MChem., MA (Cantab.) and successor to St. Peter, called for an urgent study of the *'Words of God',* in order to convey his concern, as Professor of Atmospheric Science and leading Theologian, about the cascading ***'Crisis of Climate Warming',*** and the evident disarray witnessed at the *'World Conference:',* summoned to address this matter in Paris (2015).

The Pope has attracted criticism for being preoccupied with the human condition at the grass-roots level, *'The ordinary people-as representavive of a basic point of view',* concerning the *'realities of geographical existence'* they were experiencing, by scientists lacking a sense of ecological care for those they regard as lowly cultured; as well as by senior clerics leading comfortable lives *'far from the madding crowd'.* In this context, Pope Francis is closer to St.James' definition of true religious practice, ***('Accept and submit to the word that has been planted within you and can save your souls. <u>But you must do what the word tells you, and not just listen to it and deceive yourselves.</u> Pure unspoiled religion, in the eyes of God our Father is this: coming to the help of orphans and widows when they need it, and keeping oneself uncontaminated by the world'):*** *Letter of St.James.* **(25).**

Could it be just another coincidence that the Pope at this time happens to be a scientist of the first order, as well as the world's leading Theologian, just when needed? Like Charles Darwin, the great Naturalist, well equipped to know, what is perceived at the emergent stage in any system (Human Organization, Geophysical Parameters, Combinatory Life Processes, *'Carbon cycle of Oxygenation by Photosynthesis'*), may be understood as shaped by previous events and, as the harbinger of things to come. **That is a different concept from Laws of Regulation, such as *'The Law of Constants'*, which determines the colour of the spectrum you see, related to the wavelength of light, every time a photon strikes a piece of matter bearing the particular spectrum colour identity, associated with that element. Richard Feynman** (1918-1988), recognized as a leading pioneer in formulating *'Quantum Theory', explaining the interactions between electrons and electromagnetic fields'.*

Quantum Electrodynamics: 'Quantum Fields' could appear to be acting as particles (photons), as well as waves. Einstein imagined being fixed to a photon, when pondering the relationship between interactions and events at varying time scales, and different magnitudes of perception (Relativity). The tendency in the Human Mind, confronting interactions and events at different scales, is to assume that connections at very small scales of perception, (microscale and nanoscale, one billionth of a metre), should be relatively simpler to understand.

His astonishing discovery was, systemic organized

complexity exists at every scale of perception: *('It bothers me that, according to the laws as we know them today, it takes a computing machine an infinite number of logical operations to figure out what goes on in no matter how tiny a region of space, and no matter how tiny a region of time').* He was totally perplexed to find that quantities, like the mass of a particle or atomic weight, which Physicists had taken for granted as fixed and unchangeable, were not at all immutably fixed laws. Instead, they varied, depending upon the scale from which they were viewed. *('We do not have a good* *mathematical way to describe the Theory of Quantum Electrodynamics'):* Richard Feynman: Quoted by Michael Brooks: '13 Things that don't make sense': Ch. 3. **(26).** Concerning the relation of Physics to other Sciences, Richard Feynman noted in one of his brilliant educational dialogues, that *'there is a problem in sister sciences, such as Geology, Chemistry, Biology, Meteorology & Psychology, which does not exist in Physics: viz. The historical question, "how did it get that way?".* **If we understand all about Biology** (and of course, Geography, th0e 'Queen of Sciences'), **we will want to know how all the things which are on Earth got there: e.g.** 'The Theory of Evolution: 'How the entire Earth was formed in the beginning? What kind of matter there was in the world? How did the Stars evolve, etc.?*

We do not imagine at the moment (April,1988), **that the laws of Physics are somehow changing with time—Of course they may be, and the moment we find they are, the historical question of Physics will be wrapped up with the rest of the history of the Universe'):** 'Essentials of Physics: Six Easy Pieces': Ch.3. pp.47-67: Richard Feynman; Theoretical Physicist & Educator. **(27).**

In this context, he discusses the *'possible relation between gravitation and the force of electricity',* pointing out that there is no explanation at present of gravity in terms of other forces, such as the force of electricity between 2 charged objects, *'which is a Constant'.* He emphasizes that, *'although the laws governing the 2 forces involve the same function of distance, the so-called "Unified Field Theory' may be regarded as only a very elegant attempt to combine them".*

The Time/Matter illusion revealed by Quantum (Qualitative) Electrodynamics: Scientists, who once were apparently so sure that they were mentally equal to contemplate the *'Mind of God',* the Creator of all things on Earth, now concede albeit reluctantly, that nothing in the

Universe can be considered as solid or fixed. All that is perceived, *'the reality of Geographical existence and experience'*, is now known to be changing, in a constant state of flux-including the macro-molecular assembly that **underpins the astonishing Cognitive Processes of the Human Mind.** Thus, both the concepts of *'Time & Matter'*, are revealed as illusory and misleading conceits. Yet some still remain seduced by this delusive and disedifying state of mind, in pursuit of wasteful and exploitative projects, such as *'mining on the Moon', & 'establishing a colony on Mars'.*

Who would be sent to shorten their lives, by carrying out such dangerous and costly experiments? Certainly not the highly paid eccentrics who are willing to devote their lives to such thoughtless follies, regardless of their fellow human beings here on Earth, lacking in the most basic needs, such as clean drinking water. Also, even more nonsensical ventures, such as searching for the conditions which existed in the Cosmos, *'before the Big Bang, 13,500 billion light years ago'*, shows how vulnerable the Human Mind is to being led astray by misleading fantasies, based on shallow scientific assumptions. In that way, erroneous thoughts may easily take root and spread, with potentially dystopian effects. It is salutary to keep in mind the essential message transmitted by *'QED: Quantum (Qualitative) Electrodynamics'*, that what emerges must be understood as the highest probability, shaped by conditions, and associated human behaviour of all that has preceded such events as, *'Covid 19 Pandemic' & Climate Warming'*. In this connection, it would be wise to pay heed to a scientist of distinction: e.g. **('We do not know how the Universe got started, & we have never made experiments which check our ideas of <u>space and time</u> accurately, below some tiny distance'): '6 Easy Pieces':** *Ch.2: 'Basic Physics': pp.23-45: Richard Feynman.* **(28).**

Certainly, there is no time to waste in activities based on idle speculative and fanciful assumptions. There can be no doubt that our world home is exhibiting signs of turbulence and increasing malaise, which threatens to become unmanageable. There is urgent need therefore, to harness all human resources of intellect, including scientific prowess and Theological wisdom, in order to counter this impending crisis of pernicious disorder. Can it be just another coincidence that, when *'world systems'* appear to be on the point of breaking into a state of chaotic turbulence, combined with the hateful effects of rumbling discontent and resentment arising from the unsustainable *'inequality gap'* between the poorest and ultra-rich fractions of society, renowned *'Educative Minds'* often seem to emerge? For example, Plato was inspired to establish his *'Academy'*, for the purpose of correcting disruptive erroneous concepts; Thomas Aquinas came into prominence n the 13th Century, at a time of mounting

turmoil, with the onset of the ravages of murderous Mongolian hordes, followed by the fearful *'Bubonic Plague Pandemic'*, resulting in the obliteration of over a third of the world's population at that time. His great work was concerned with establishing that *'**Reason comes from the Mind of God'**,* meaning that there is a *'Principle of Order'* governing all things to their appointed end: *(**'The greatest kindness one can render to any man consists in leading him from error to truth': 'Love takes over where knowledge leaves off'**):* Thus, he can be credited with establishing a *'Philosophy of Ethics and Ecological Care'.* He is remembered as an intellectual *'Channel of Peace'*, at a time of extreme unrest, in sore need of a calming and healing influence. **(29).**

In this way he is remembered as reflecting the priestly presence of Christ, who stressed both the unity of valid knowledge, and the overriding power of a charismatic sense of *'**Love', by which the corrosive effects of hateful conflict may be sublimated,*** replacing despair with hope and an optimistic outlook towards the world. *(**'A man can have no greater love for another than to lay down his life for his*** *friends'): John 15. 13.* The few empathetic souls prepared to listen and respond to Christ's words continue to shine like a beacon of hope to all living in poverty and hopeless misery, throughout the world today. **St. Vincent de Paul (1581-1660),** a gentle and truly religious priest, was horrified by the plight of so many poor wretched people, reduced to a skeletal state by prolonged hunger, thronging the slum alleys of Paris, like wild animals ready to pounce on any crumb or scrap of food that might be discarded. With a heart full of pity, for all the misery and suffering festering around him, he was inspired to dedicate his life to do all he could to alleviate the sufferings of those burdened by unbearable hardship, struggling just to survive. Working day and night with unremitting zeal, he was greatly loved in return, and remains renowned for his compassion, humility and generosity. At the same time, the Royal Household and Court, attended by fawning Prelates and ruling politicians, were living lives of luxury in their country estates, such as Versailles, without any regard for their people condemned to miserable lives of dire poverty. A century later, the hatred engendered by this hideous state of inequality inflamed the horrors of The French Revolution, with the mass execution by guillotine of aristocrats, together with the King and Queen, and well-off professionals like Lavoisier, the famous Chemist.

Whereas *Love* begets a loving response, *hate* not only provokes revenge, but also threatens to flare up to even more destructive and vindictive level, such as the bestial behaviour of a diabolical nature, witnessed during the 2nd World War. In that way, we may deduce that patterns of human behaviour recorded and remembered in some way, from however long ago, influence and have a bearing upon the collective perception of the world and other people, as well as the behavioural response we currently observe. Inevitably therefore, present patterns of human behaviour may be expected to presage and influence the pattern of locational behaviour, to emerge in future. In this respect, it is important to keep in mind that the concepts of *'Quantity and Pattern'*, are different logical types of understanding, as demonstrated by **Norbert Wiener** (1894-1964), and **Gregory Bateson** (1904-1980), who were both seconded to the military research unit formed from various disciplines, including mathematicians, neuroscientists, social scientists and engineers, at Bletchley Park. Their specific task was to find a way of breaking the *Nazi Enigma Code,* a seemingly impossible problem involving over 300 trillion combination probabilities, far beyond computation at that time. Led by Alan Turing, the Polymath of AI *(Artificial Intelligence),* they succeeded in deciphering the code, thereby shortening the war, and saving many millions of lives. Wiener is credited as the pioneer of the new science of **'Cybernetics'**, *as an exact discipline of Mind'* , focused on studying how the human mind functions, both as a system of control, and also how vulnerable it is to conditioning, and to being subjected to control by influencing **thought patterns of conceptualization.** But patterns cannot be measured or thought of as material entities such as particles or solid matter. **(30).**

However, they can be assessed as flows and gradients of energy and, also as *'wave patterns'* of power. In that context, *'Cognitive processes',* which are part of *'Life Processes',* must be understood as knowledge-based systems of communication. Recognized as among the leading pioneers in establishing the developing science of Cybernetics and Logic of Mind, Gregory Bateson defined Consciousness as a fundamental system of organization and regulation: *('Operating like Medicine in its sampling of the events and processes of the body , and of what goes on in the total Mind-- It is organized in terms of purpose (teleonomy)--a short-cut device to enable you to get quickly at what you want; not to act with maximum wisdom in order to live, but to follow the shortest logical or causal path to what you want next. Above all, it may be money, or power'): 'Steps to an Ecology of Mind': 'Conscious Purpose versus Nature'.* (**31**).

Norman Wiener and Gregory Bateson became close friends, deriving enhanced power from each other's mind, which enabled them to establish a systems approach to a range of phenomena, including family therapy, alcoholism, schizophrenia, and mental disorder leading to learning impairments relating to language processing, computation skills, memory, attention deficit, and social behaviour. In that way, they opened a renewed understanding of the capacity inherent in the Human Mind, doing away with the Cartesian division between mind and body, the false methodology which has had the effect of nullifying scientific inquiry for centuries by simplifying and reducing complex dynamic systems to mere mechanisms. *('The social system is an organization like the individual, that is bound together by a system of communication, --and it has a dynamics, in which circular processes of a feedback nature play an important role'): Wiener, N.: Quoted in 'The Web of Life': Fritjof Capra.* (**32**).

In connection with these wartime code-breaking activities, Wiener noted, whilst developing a computer model to simulate mental activity in *'Cognitive Processes'*, that it was a false assumption, *'to reduce cognition (the process of "knowing"), in the human mind to a mere mechanism of "information processing", viewed as a manipulation of symbols according to a set of rules.* Thus, he was prescient in foreseeing that this new area of psychological research, *'Artificial Intelligence (AI)'*, was not only an absurd reduction of the extraordinary cognitive powers of the human mind, with capacity to experience emotions -such as **love, truth, empathy, imagination, etc.,** but also was a delusion fraught with dire consequences.

He warned that this erroneous train of thought could have the effect of subverting the nascent science of *'Cybernetics'*, **based on the study of Logic of Mind & Theories of Learning, thereby reducing a human being to the status of a robotic entity.** In 2018, robotic machines were replacing human labour widely in the workplace, and Russia had announced the production of a robotic nuclear armed submarine, capable of operating devoid of human crew. *('Conscious Purpose is now empowered to upset the balances of the body (human metabolism), of society, and of the biological world around us, threatening a loss of balance. Lack of systemic wisdom is always punished— Biological systems—the individual—the culture and ecology are living sustainers of their component cells as organisms—but the systems are nonetheless punishing of any species unwise enough to quarrel with its Ecology. Call the systemic forces 'The mind of God' if you will'----This massive aggregation of threats to man and our ecological systems, arises out of errors in our habits of thought at deep and partly unconscious levels—We must achieve clarity in ourselves, then look for every sign of it in others—and reinforce them in whatever is sane in them'):* Gregory Bateson: *'Steps to an Ecology of Mind'.* **(33).** When renowned logical minds of this calibre, who have also demonstrated outstanding *'ontological wisdom in order to live',* converge in expressing their concern about mounting disorder in world systems, it is time for all to sit up and take notice. In this respect, the collective ecological insight which opens comprehension of *'Life and the World',* scientifically or theologically, must be regarded as a **phenomenon of educational communication,** opening the mysterious capacity of consciousness in the human mind to a higher level of understanding.

The Central Problem of the Human Mind as a Quantum Field of Love: Richard Feynman drew attention to the

'Physiology of Sensation'. ('The central problem of the Mind—or nervous system is, what happens when an animal learns something—it can do something different than it could do before—and its brain cell, made of atoms (molecular systems), must have changed too. We do not know where to look, or what to look for when something is memorized. We do not know what it means, or what change there is in the nervous system when a fact (knowledge) is learned. The brain is such an enormous mass (network) of interconnecting wires and nerves, that it probably cannot be analysed in a straight forward manner'): Ch.3: 'The Relation of Physics to Other Sciences': p. 64. **(34).**

Philosophy of Immanence: Search for Wisdom inherent in the human Mind: This concept, stems from a simple recognition and acceptance, inherent in the natural mind, and reinforced by geographical experience, that there is a good way of thinking, behaving, and acting towards fellow men, which ensures harmonious relationships, and leads to a propitious way of life. That is a rather different way of looking at the complex *'Web of Life',* defined by Richard Feynman as the *'Central Problem of the Human Mind',* which seems to be more concerned with the nuts and bolts of consciousness, and how the mind works. That would be a mechanistic view of Life and the world, which rests on the assumption that knowledge concerning the parts and connections of a complex network of interactions, summed up, will provide full understanding of the whole. But *'Quantum Electrodynamics'* gives a different message. If the accretion of energy in world systems, resulting from the incremental kinetic power of the *'Photo-Electric Effect'* is taken into account, it becomes obvious that full understanding of the *'Life-World Enigma',* is beyond the capacity of the human mind. *('No attempt to explain the world scientifically or theologically can be considered to be successful unless it accounts for the paradoxical conjunction of <u>"being and</u>*

becoming". **Can we regard the countless atomic processes going on quite naturally, all the processes inside you and me, inside the stars, the interstellar gas, distant galaxies, as some sort of gigantic cosmic computation? If so, we would arrive at an astonishing conclusion: the Universe would be its own stimulation!'):** *'The Goldilocks Enigma': & 'The Mind of God'.***(35).** That observation carries a profound implication: the human mind must be regarded as a **phenomenon of communication, conferred by a greater Mind, imbued with the *'organization of becoming'*,** by which all things are connected.

A Universal Principle of Order: Furthermore, Sir Roger Penrose (Mathematician & Quantum Theorist), points out that the human mind is part and parcel of of this *'Cognitive Process of Knowing',* and not to be considered as outside of this astonishing emerging network of *'ordered dynamic interactions',* as if looking on. *('The faculty of human understanding is beyond any computational scheme whatever: Consciousness is part of our Universe, so any physical theory which makes no place for it fails fundamentally short of providing a genuine description of the world-- There is yet no computational or physical theory that comes close to explaining our consciousness and consequent intelligence'):* *'Shadows of the Mind'.* **(36).**

Contemplation of the *'Learning Process'* and thought development in this way, opens awareness of ourselves, enabling the ability to see ourselves as others see us.

('Oh the gift that God could give us, to see ourselves as others see us'): *Robert Burns: 1786.* Ancient Greek Philosophers put it more succinctly: *'Know yourself'!* Clearly, the cognitive faculty of feeling and *'Sensation',* defined as the capacity to think sensibly and with care, must be regarded as a higher order of awareness, as proclaimed in ancient Judaic Scriptures: *('Wisdom is bright, and does not*

grow dim, By those who love her she is readily seen, and found by those who look for her. Quick to anticipate those who desire her, she makes herself known to them. Watch for her early and you will have no trouble- you will find her sitting at your gates. Even to think about her, is <u>understanding fully grown;</u> be on the alert for her and anxiety will quickly leave you. She herself walks about looking for those who are worthy of her, and graciously shows herself to them as they go, in every thought of theirs coming to meet them. Of her, the most sure beginning is the desire for discipline. Care for discipline means loving her'): Wisdom 12-29.

Thus, we can begin to see the world in a very different way, and are meant to do so, including every living organism, as an integrated whole, rather than a dissociated collection of parts. *('That is a deep ecological view recognizing the fundamental interdependence of all phenomena. Whether as individuals, or societies, we are all embedded and ultimately dependent on the cyclical processes of nature'):* Capra, F: 'The Web of Life'. **(37).**

That mode of thought opens appreciation of our own being, knowing that we are both a product and part of a wondrous pattern of organization, by which Life, our world and the Universe came into existence. The same fount of enlightenment you will find, empowers a strong sense of appreciation for our Earth-world home, springing from a well of empathy, and commanding care for our family, society relationships, our country, world order, and an improving ambience for our offspring in the future. The very concept of a disciplined way of life reflects a natural desire for well-being, prosperity, and especially orderly patterns of resolution in all the trials and tribulations that may unfold on Life's journey, such as those which beset our present world: *'Crisis of Climate Change'; 'Pandemic Disease'* & smouldering discontent coupled with looming conflict. No wonder that the beauty of our Earth-world home, prepared and ordered, as we have found, to be just right for Life to come into being and flourish, has become obscured for so many people by a deadly mood of pessimism, clouded with hopelessness about the future and the belief that everything turns towards evil.

The younger generation particularly, have been afflicted by this sinister *'spiritual pandemic of nihilism'*, sweeping through human relationships across the world, convinced that life has little to offer, as shown by the increase in suicides. There is an urgent need to counter this pernicious influence, which also has become a divisive effect between the younger generation and their elders, when the older generation are blamed for all the harm done to the world in the past, viewed as an additional intolerable burden of care. It becomes imperative to remind ourselves that the message of *'The Information Age'*, tells us -as part of the world state, that we have also the capacity, *'wave-energy of cognitive power'*, to pull disorder back to an **innate state of order.**

('The Process of Life is the activity involved in the continual embodiment of the system's pattern of organization. In a living organism the pattern of organization is always embodied in the structure of the organism, and the link between pattern and structure lies in the process of self-embodiment. The definitions of the 3 criteria, which are all totally independent, are 3 different but inseparable perspectives of the phenomenon of Life: 1) Organization, relationships that determine the essential characteristics of a system; 2) Structure, the physical embodiment of the system's pattern of organization; 3) Life Process, the activity by which the embodiment of the system's pattern of organization unfolds and emerges'): Erwin Schrodinger: 'Order from Order' :1943. **(38).** That observation is a profound revelation, worthy of thought. Spiritual awareness springs from the opening of mind to comprehend the phenomenon of *'being'*, to value Life as precious, and know

yourself as a valued creation of an omnipotent caring '*Mind*'. That knowledge-based, systems-view of Life, engenders the spiritual qualities of optimism, hope and care, which fountain into '*A Life-spring of Faith*', defined as the **bond of Loving care: (Deep Ecology).**

('On the way He came to the Samaritan town called Sychar—Jacob's well is there, and Jesus, tired by the journey, sat straight down by the well. It was about the 6th hour when a Samaritan woman came to draw water. Jesus said to her, "Give me a drink!". His disciples had gone into the town to buy food. She said to him, "What? You are a Jew, and ask me, a Samaritan, for a drink?" Jews in fact, do not associate with Samaritans. Jesus replied: "If you only knew what God is offering, and who it is that is saying to you, "Give me a drink, you would have been the one to ask, and He would have given you Living Water—welling up to eternal life"): John 4. 5-22. (**39**).

Where Science, Education & Spirituality meet in unity: Plato came to the conclusion that water, must be the prerequisite of *Life*. Thence, the formula, H^2O. Initial revelations concerning '*Quantum-electrodynamics*', and specifically the '*Quantum behaviour of Water & the Interactions of Quantum Fields*', suspected that the '*Life support system of Photo- synthesis*' (oxygen produced by energy of photons of light), involved splitting carbon molecules, but then it was discovered that water molecules were being split by sunlight in the '*oxygen evolution cells*' of leaves to produce the oxygen essential for life to exist, with carbon dioxide as a waste product, also necessary for plant metabolism. (**40**).

Quantum Behaviour: We may learn from this advance in knowledge about water, that it would be very rash to assume that a little knowledge about the interactions, linking Life systems with Earth's extremely complex web of geophysical

65

systems, could provide an explanation of the *'atomic behaviour'* involved: (Human Organization<-->Environmental Systems)

('Because atomic behaviour is so unlike ordinary experience, it is very difficult to get used to-- it appears peculiar and mysterious to everyone, both to the novice and the experienced Physicist. Even the experts do not understand it the way they would like to, and it is perfectly reasonable that they should not, because all direct human experience applies to large objects. We know how large objects will act, but things on a small scale do not act in that way. So we have to learn about them in a sort of abstract or imaginative fashion and not by connection with our direct experience.): *'Essentials in Physics': Ch.6:'Quantum Behaviour': pp.115-138: Explained by Richard P. Feynman, acknowledged as a most brilliant Teacher.* **(41).**

Note these implications:

- *Both Life and our Earth-world are revealed as masterpieces of self-organization, with all parts and connections working as patterns of coordination, in support of vital systems.*

- *Life systems, endowed with mystical qualities of meaning and purpose are not amenable to analysis by reduction to the state of mechanical subsystems, and lumping all the bits together, on the presumption that the complex network of fluid dynamic interactions can be explained and understood in its entirety. To understand just one subsystem in the chain, it would be necessary to understand all the links in the chain.*

- *'Thus arose the tension between wholeness and fragmentation between body and soul which has been identified as the essence of the human condition by*

poets, philosophers and mystics throughout the ages':
Fritjof Capra: 'The Web of Life'. **(42).**

- *The Systems view, offers to point the way, in mitigating and resolving problems of turbulence and disorder, and especially to restore harmonious relations between hostile human factions. But this would depend on whether the Human Mind remains open and amenable (educable), to perceive and process valid (epistemological) Knowledge.*

- *Therefore, erroneous thoughts propagated in the past, must be identified and eradicated, including misinformation, false and misleading interpretations, & insidious cults and addictive ideologies, such as virtual-world experience/ gaming/gambling/AI control of mind.*

Universal Order: Therefore, the key challenge, facing a world in a state of turbulence and impending disorder, becomes an urgent educational problem, the need to develop the capacity to think in a *'qualitative'*, rather than a *'quantitative'* frame of mind. *('**Mind and consciousness are not things, but cognitive processes, identified with the very process of Life, at all levels, revealing two complementary aspects- its process and its structure')*: Capra, F. & Luisi, P. L:'The 'Systems View of Life': CH. 17-- 2. 1.* **(43).**

With the ability to think in this enlightened way, at world level of governance, we may know that all phenomena, simply observed, are to be understood as emerging and coming into existence in a chain of organized contingent connectivity. Not predetermined, but from systems in a constant state of fluid dynamic flux, yet in accord with an innate state of order. For that reason, and to reawaken a sense of spiritual renewal, in 2019 Pope Francis, *Theologian,*

Scientist & Educator, announced the inauguration of two new *Feasts: 'The Word of God', and 'St. John Henry Newman',* to commence in February 2020. It must be regarded as profoundly significant, that both celebrations hark back to a seminal educative ethos: *'The Word of God',* from ancient Judaic Scriptures, as encapsulated in *'The Psalms'* and *'Ecclesiastes'* in particular, rather than generally in *'The Old Testament',* which is full of errors.

'Cardinal John Henry Newman' (1801-1890), together with *'Anglican Bishop Edward Pusey'* (1800-1882), were leaders of *'The Oxford Movement',* dedicated to the restoration of *'the Priestly presence of Christ amidst the Catholic assembly'* of the Anglican Church, a longed for revival, alternative to the spurious *'Church of England',* set up by Henry VIII to satisfy his own psychopathic lusts. Bishop Pusey, Professor of Hebrew in Oxford University, deserves recognition as among the most erudite and spiritually inspired Prelates, as shown by his demonstration of **faith in 'Christ as the 'Saviour of the World':** e.g., his sermon to the University assembly in 1843: *'The Holy Eucharist, and desire for the priestly presence of Jesus Christ'.* **(44).**

It cannot be dismissed as just another coincidence, that when conditions in the world take what appear to be yet another cyclical turn for the worst, fraught with moral depravity, with even worse impending disaster, the greatest educative minds emerge, just when needed. At the beginning of the 19th century, the European world was left at a nadir of disorder, wretched and impoverished, as a result of hateful conflicts between the ruling monarchies, culminating in the wasteful Napoleonic campaigns. Britain was left tarnished by the obscene cruelties inflicted upon Ireland, on a genocidal scale, by which the population had been reduced by the death of 2 million during the Great Famine *('An Gorta Mor:1845-1852),* and by the forced migration of another 2 million people, under the despotic rule of an *'Anglo-Irish Protestant minority'* administration, set up for the purpose of supplanting *'The Catholic Assembly in Ireland',*

by a spurious *'Church of Ireland',* with the intention of forcing the people into servile submission under British colonial rule. Many young men, apprehended as they met at crossroads, on the way home from work in the fields, were sentenced to be transported to the colonies, for life with hard labour imprisoned for disobeying draconian laws forbidding such meetings. Ironically, because of these penal laws, it has been estimated that there are now over 125 million people of Irish and Scottish descent, in Australia and America, and scattered across the world. Yet, at the same time as Europe, including the British Isles, appeared to be collapsing into a state of seemingly irremediable turbulence and distress, **Bishop Pusey and Cardinal John Henry Newman** came into prominence as *'Great Educators'.* For example, in 1852 Cardinal Newman published his classic book, leading to the foundation of Dublin University0. **(45).**

He had been requested to undertake the task of redressing the ruthless subjection of the age-old Catholic cultural ethos of the Irish people, which had commenced under the psychopathic rule of the Tudor Monarchy, in the 16th Century. The harsh penal laws imposed on the Irish population, included the suppression of their ancient rich heritage of the Gaelic Language, the Celtic branch of Indo-European Languages spoken by the Galatians, mentioned in the Letters of St. Paul. In order to enforce this merciless policy of *'Ethical* cleansing', a new University College was established in 1592, (*'Trinity College of The Holy Undivided Trinity of Queen Elizabeth'*), located in the dispossessed old *'Augustinian Priory of All Hallows',* for the purpose of consolidating the supremacy of, *'Protestant Ascendancy',* instigated under the despotic rule of Henry VIII.

Students adhering to the *'Catholic Faith'*, were barred from entry. You may recognize parallel circumstances unfolding in our world today. Extreme psychopathic behaviour of this nature, showing inability to feel pity or care for the sufferings of fellow human beings, is symptomatic of mental disorder, in which the spiritual cognitive quality of care and bonding *'sensation of love'*, naturally endowed in the human mind, becomes negated, supplanted by the deadly demonic influence of *'hatred'*, which even finds pleasure in killing and inflicting torture on anyone who stands in the way. However, *'The qualitative power of Truth and Love'*, viewed as immutable and eternal laws inherent from the moment of creation of the Universe, cannot be nullified, as Christ assured in the *'Beatitudes'*: *('Blessed are the humble (poor in spirit) for they shall see heaven; Blessed are the gentle for they shall inherit the Earth; Blessed are those who mourn for they shall be comforted; Blessed are they who hunger and thirst for right for they shall be satisfied; Blessed are the merciful for they shall be shown mercy; Blessed are the pure in heart (with the innocence and simplicity of a child), for they shall see God; Blessed are the peacemakers for they are the sons of God: Blessed are the persecuted for theirs is the Kingdom of Heaven'):(46)*

The proclamation of those spiritual qualities, as *'Blessings'* conferred by *'The Word of God'*, in the *'Sermon on the Mount'*, confirms a unity *of mind and bodily behaviour, showing how perceptions and actions can be influenced by evaluations that emerge from aspects of geographical experience'. ('Yet properly presented, Quantum Theory is thoroughly in line with our deep human intuitions. It is the 300 years of indoctrination with basically false ideas about how nature works that now makes puzzling, a process (cognitive), that is completely in line with normal human intuition'): 'Information and the Nature of Reality' Ch.6.*

'Minds & Values in the Quantum (Qualitative) Universe':
Henry Staff. **(47).**

In this context, Cardinal Newman was keenly aware that hateful disordered minds like the Tudors and their henchmen, are bound to generate further and even more insidious turbulence and disorder in systems of human organization, such as the Anglican Protestant network of complicity, which can continue to act as a corrosive influence in human relationships, until purged in some way. That is how he saw the world around him in Britain, as it was at the turn of the 19th century, in a decrepit state of dilapidation. With the onset of *'The Industrial Revolution',* rural patterns of relationship were collapsing, as the younger generation left in droves, drawn by the lure of higher wages available for work as industrial labourers, in squalid housing conditions spreading like an urban rash, ill-planned and hastily built on the outskirts of towns. These cramped and insalubrious environmental conditions provided the ideal conditions for the incubation and proliferation of pandemic disease, such as cholera, typhoid, and tuberculosis.

Consequentially, the deathrate increased dramatically, leaving many orphaned children living in conditions of dire poverty, eventually consigned to a *'Workhouse'* under the supervision of a *'Parish Officer',* to be prepared and sold-off as cheap slave labour. In this way, Life and human beings can be devalued, to the point of being regarded as disposable and of little worth. Charles Dickens (1812-1870), has left a poignant record of this tristful train of events, showing how children were treated in the *'Workhouse',* during the Victorian government regime: e.g., *'Oliver Twist'.* When the political ruling class find it expedient to make use of the *'Church',* as a cloak of respectability to justify such unchristian treatment of the waifs and strays of society, the tangled web of deceit tends to spread as a self-perpetuating cankerous influence. In that way the political aristocracy have been able to acquire great wealth under a cover of moral rectitude, whilst leading churchmen were persuaded, by the award of sinecures involving little or no work, to remain passive silent accomplices, by ignoring the plight of the down-trodden poor, such as the *'Workhouse children'* working as slave-labour, under Victorian rule.

In Ireland the anti-Catholic penal laws, supported by *'The Church of Ireland'*, even took advantage of the extreme poverty during *'The Great Starvation'*, by encouraging children to report their parents and relatives for adhering to the Catholic Faith, by rewarding them with the family confiscated possessions. In the colonial penal settlements set up in Australia, such as Norfolk Island, Tasmania, *('Van Diemans Land'- known as the "End of the World")*, & *'Botany Bay'*, the harsh treatment imposed upon transported convicts was so brutal that it makes harrowing reading: e.g., the 200 children sent to *'Point Pure-Tasmania'*, in 1842. 165,000 convicts were transported to Australia during the 18th & 19th centuries: *'The Fatal Shore': Robert Hughes.* **(48).**

One can scarcely imagine how grim and ugly the world must have appeared to a little girl, condemned to work long hours in a woollen mill, as slave labour during the 19th century; or to a little boy of 7, forced to climb inside chimneys as a chimney-sweep, bald and suffering from acidaemia; or to the little girl of 11, transported to Australia in the First Fleet (1786), for stealing a ribbon in a Fair. A world perceived as bereft of the *'Qualitative healing sense of loving care'*, engenders fear and the pessimistic sense of despair, where *Life* seems to be *'not worth living'*, the very antithesis of the world proclaimed in ancient Judaic Scriptures, as *Created by the Word of God: ('In the Word there is Life, and Life is the expression of the Word').* Furthermore, lack of concern for human life, and the human conditions, is conducive to a *'devil-may-care'* attitude towards the effects of human exploitive activities upon the quality of our precious

'Biosphere', now revealed as *'a mysterious Quantum Field of Life-Support': Moran, J..J: 'Know Your World': Ch.3: 'Ethereal Atmosphere': & Ch.4: 'The Thin Blue Line: Care of the Biosphere'.* **(49).**

Quantum Field of Univers0al Order: In this respect, Pope Francis together with pioneer scientists concerned with *'Life Processes, especially the Biogeographical Disciplines',* has focused attention upon the message conveyed both by epistemological knowledge contained in theological teaching, explained by Christ, and reinforced by the revelations of *'Quantum (Qualitative) Electro dynamic Theory.* The challenge is to study the meaning and significance of the emergence of Life, as an *'Ordered System of Creation'.*

Life Processes of all created things, according to the creative words of God: (*'Let us make man in our own image, in the likeness of ourselves'...'be fruitful and multiply, fill the earth and conquer it'...thus heaven and earth were completed with all their array'):* Genesis 1 &2. This account, written by learned scribes long ago, implies that we are required to take care of our home world, as well as of each other. That our planet earth has the exact gravity-bound environmental conditions, formed as a *'cradle to support Life',* must be regarded as a profound mystery and not just as a matter of coincidence: *('The Goldilock's Enigma': Paul Davies).* Furthermore, the realisation that, *'all living beings share the same world together, which should enable us to bring forth an improving world environment for the good of all',* as well as inspire us with the deepest ecological desire to take care of it, especially our neighbours and all living things. *('The Systems View of the World: A Unifying Vision': Capra, F. & Luisi, P.L.).*

How have we responded to this challenge? The answer is only too plain to see. The habitable environment bears all the signs of severe degradation, following reckless destructive human activities, especially the ravages of barbaric wars. Systems of human organization are tottering on the brink of unmanageable disorder. Society relationships are souring, leading to the formation of hostile factions prepared to resort to hateful criminal activities against each other. The very meaning of *'neighbour', (Saxon: nachbur= nearest farmer, hamlet, village),* the family group you could rely upon for help when needed, or to seek a wife or husband in times gone by, has been lost. The bonds of mutual brotherhood, even in the family group upon which healthy society relationships are based, have become weakened: e.g., by seemingly anarchistic movements, as provoking a mood of resentment amongst the younger generation against their elders, by encouraging schoolchildren to rise as a rebellious mob, pointing the finger of blame at the older generation for having caused the menacing 'crisis of climate change': (*'We shall never forgive you for robbing us of our childhood': Time Magazine, 23 Sept. 2019: 'The Greta Generation').* Ironically, this movement originated in the place which gained notoriety as *'The Merchant of Death',* the major manufacturer under the guise of neutrality, of high-grade steel and the vast quantity of explosives required to supply the barbaric wars of the 20th Century. The great wealth acquired has left them amongst the richest in the world, but also noted for having a dismally materialistic outlook.

The wounds inflicted on the world environment, by those hateful conflicts, as well as the associated divisive effects upon society relationships, continue to fester. No longer can we face our Creator God, with confidence, that we have fulfilled his fatherly admonition, (*'to love our neighbour as ourselves, as well as to take good custodial care of our home-world': ('A house divided against itself cannot stand': Mk.3.25: & 'It is not those who say to me, "Lord, Lord", who will enter the kingdom of heaven, but the person who does the will of my father in heaven': & 'Therefore, everyone who listens to these words of mine and acts on them will be like a sensible man who built his house on rock':* Matthew 7. 21- 27.

No wonder that the younger generation are showing signs of being engulfed by a sinister cloud of pessimism that is sweeping through the world, with the insidious effect of closing down the higher spiritual cognitive qualities of mind, even to the point of losing the will to live. In this way, it is evident that the forces of darkness have promoted an ethos of agnosticism and atheistic denial of the existence of God, thus reducing the **'Word of God'** to mythical thinking from long ago, promoting the belief that we live in *'a secular society'*, where we are free to do as we like: the nihilistic philosophy of, *'after all, if everyone puts themselves first, why shouldn't we?'* This view of the world, as a dangerous environment, and your neighbour as an adversary not to be trusted, is not conducive to a sense of optimism, but tends rather to promote an outlook of fear and dread about what may happen in future. Thus, the higher spiritual qualities of cognitive awareness, which induce a sense of happiness and well-being, **'Knowledge, Faith, & Caring Qualities of Mind'**, may be blunted or shut down.

('This is a very different way of viewing the world from that advocated by classical scientific thought, especially from physicists. For a start it springs from a sense of awareness that all we perceive in the interplay between world systems (human priorities linked with geophysical parameters), is coloured and influenced by an understanding that the human mind is itself an essential part of the Life systems which we are seeking to analyse. Indeed, the capacities of the human mind are to be seen as an endowment for that very purpose: i.e. Cognitive Processes are part of Life Processes, and cannot be separated'), 'Applied Geography': Ch.3: 'Reawakening of Spiritual Qualities of Mind': Moran, J.J. **(50)** .

These thoughts, prominent in recent homilies, have inspired a renewed analytical consideration, concerning *'The Word and Will of God':* viz : -

18 *Nov. 2018: 'Heaven and earth will pass away, but my words will not pass away': Mk 13.31*

4 Aug.2019:'Man's life is not made secure by what he owns, even more than he needs :L13.31

1Dec.2019:'Therefore you must stand ready': Mat24.4: 'Society drifts back to a 'Secular World'

26Jan.2020:'Repent for the kingdom of Heaven is close at hand': Mat 4.17 : 'Dialogue with God'

2Feb.2020:'Lectio Divina'::The Word was the true light that enlightens all men': John 1: 'God is not remote, but at our side to console and encourage us.'

1March 2020:'Man does not live on bread alone, but on every word that comes from the mouth of God': Mat 4.4: The Human Mind has been endowed with the cognitive capacity to enable us to communicate with the Holy Mind of God, empowered by the mysterious faculty of *'Language',* if only we have the faith to believe it, through the spiritual qualities of Truth and Love: *'I am the light of the world; anyone who follows me will have the light of life': John 7.12.* At the same time, a stronger sense of semantic understanding seems to be unfolding. It is as if the thoughts seeded in my mind in 1936, as a pupil in London Oratory School, when studying the *Parables* in which Jesus explained the creative *'Words of God'* by stories, which even a child could understand, have taken roots and sprung into growth. There can be no doubt that our powers of analysis depend on a spiritual power of *Faith, endowed in the human mind,* capable of cultivating a philosophical outlook of peace, with a desire for harmonious relationships which unite, rather than souring conflicts, which tear nations and the world apart.

Clearly, the more we are enabled to extend our knowledge about its systemic connections, including human cognitive processes, the more our capacity of understanding is empowered to comprehend: cf. *('Knowledge is like a sphere, the greater its volume, the greater its contact with the unknown':* Blaise Pascal *)*.

Jesus made this very clear: *('Were your faith the size of a mustard seed* (speck of dust), *you could say to that mulberry tree, "Be uprooted and planted in the sea, and it would obey you':* Luke 17.5).

It follows, that in this way the higher aesthetic qualities of understanding and appreciation of all that is good, with which **God has enriched the human mind**, may become energized, together with a custodial sense of ecological care towards the well-being of our fellows, as well as the world environment. In this light, as surely as the darkness of night gives way to the brightness of day, we may know that **The Word of God is not to be perceived as a distant voice,** but rather as speaking and resonating constantly in the cognitive processes of human consciousness. It is salutary to be reminded that the mysterious faculty of *'Language'* is of profound significance *(Semantics),* as an empowerment of understanding, enabling us to communicate and convey the meaning *(Semiosis),* of all we have learned, in the light of the *'spiritual qualities of Truth and Love'.* Ecological principles do not go hand in hand with the baser motives of human nature, such as the pursuit of power and wealth, which have become dominant influences in our modern world.

Language Codes as Phenomena of Organization: The human mind, invested with the capacity to be in dialogue with the **Mind of God,** enables us to perceive ourselves in a new way, as a fundamental system of cognition, with the power to pull *'disorder back to order'.* *('Order From Order':* **'3 criteria- pattern; structure; & process, are different but**

inseparable perspectives of the "Phenomenon of Life"): Dublin Lectures: 1943: Schrodinger, E.).

In this context, we are made to appreciate the wonder of our being, through the revelation of 3 phenomenal concepts: - **Autopoiesis:** Life defined as a system capable of sustaining itself through a network of reactions, which continuously regenerate the components within a boundary of its own making. *('The Systems View of Life':Ch.12: 'Mind & Consciousness': Capra, F & Luisi, P.* **Cognition & Capacity of Mind:** Cognitive Processes are intrinsic qualities of spiritual awareness and communication which cannot be separated from Life Processes, to be regarded as precious and mysterious an endowment as *Life* itself, enabling us to know that we are experiencing the phenomenon of communication with the *'Mind of God'. ('The Mind of God': Paul Davies).* **Meaning & Purpose:** Enlightened in this way, the capacity of understanding becomes enhanced in the human mind. Thus. we are experiencing, in the 21st century, an episode of revelation so remarkable, that it has astonished scientists, leaving them bewildered and unsure about the very *'Laws of Physics' (Nature),* upon which they have relied, to explain their observations as patterns of regulation. *('If the future of science depends on the things that don't make sense, the Cosmos has a lot to offer': '13 Things that don't make sense': 'The Missing Universe': pp. 1-35). Michael Brooks).*

Convergence of scientific thinking with the teaching of Jesus Christ: The developing, *'science of control and communication, (Cybernetics)',* has brought together mathematicians, neuroscientists, engineers, social-scientists, philosophers, theologians, and all concerned in studying patterns of communication, especially as closed loops & networks of feed-back operations in systems of self-regulation and organization. Thus, the complexity presented by the dynamics in the workings of network loops, has emerged as a central concept, *'Theory of Cybernetics', the science of communication and connectivity*, based on valid knowledge.

('We are but whirlpools in a river of overflowing water—not stuff that abides, but patterns which perpetuate themselves':(Norbert Wiener (1884-1964), pioneer of *'Cybernetics', as the science of communication and connectivity between the creative 'Word of God' & the human mind.* We are receiving a glimpse of the incomprehensible complexity presented by linkage chains of molecular functions, by which Life and Geophysical systems are connected, opening our perception of Life & the World about us, to an unexpected level of comprehension, whilst knowing that no human mind can possess the capacity to understand it all: e.g. *'The Genetic Code' & 'Quantum Dynamics'.* At the same time, we begin to perceive ourselves as a vital part of it, with the capacity of an, *'integrative control system of mind'*, to restore order in our dystopian world. *('For a system to be conscious, it must integrate information in such a way that the whole generates more information than its parts...integrated information cannot be reduced into smaller components':* Giulio Tononi: Neuroscientist: 'Your Conscious Mind': Ch.2:'The Biological Basis of Consciousness': New Scientist.*(51)*

However, this sense of *'cybernetic control'* may be diminished or shut down entirely by the astringent effects of traumatic geographical experiences. Indeed, we are witnessing this pernicious influence sweeping, like a cloud of darkness throughout our world today, causing a mood of fear and anxiety to develop in the minds of the younger generation, about what may befall in the future. In this context, Pope Francis has called for a symposium, to examine the deleterious effects of gadgets and electronic devices, which are being produced for the purpose of promoting, *'virtual world & combative gaming experiences',* under the impetus of the new *'pseudoscience, AI (artificial intelligence) & Robotics'.* There is mounting evidence, that the young mind, clouded by anxiety about the future, can easily become addicted to these stupefying experiences, as an escape from feeling unloved, and what may be perceived as a dangerous world. The increasing rate of suicide among the younger generation shows that even the will to live may be lost in this way. Whilst robotic devices can be made to serve as valuable adjuncts in carrying out routine precision tasks, such as pin-point brain surgery, they can never be invested with the spiritual cognitive faculties of, *'Semantic understanding' & 'Semiotic discernment and reasoning',* endowed in the human mind for the purpose of communication with the *'Mind of our Creator'.* The fallacious premise blazoned by this pseudoscience that the world will be a richer place, and people lifted out of poverty, with unbounded time for leisure pursuits as a result of developments in *'robotic devices',* is an example of how easily *'knowledge'* can be misrepresented as biased *'information',* or subverted to *'misinformation'.* **Obscuration of 'The Word of God':** In this respect, the Bible is not exempt from censure. At school we were advised not to read the Old Testament, because it contained scandalous reading, but to trust in the teaching of Jesus, in the New Testament, for a valid explanation of: *'The Word of God',* about how we should live our lives. *('Be ye perfect, even as your Father who is in heaven is perfect': Matt. 5.48).* However, the New Testament cannot be regarded as entirely free from the transmission of erroneous concepts, which have had a harmful effect on human relationships, by propagating attitudes of despisal in place of bonds of love.

Examples of corrosive influences in human relationships: Belittlement of women: This erroneous concept had its origins in the Book of Genesis, with the myth of *'woman',* created as a companion for *'man',* which is a blatant example of non-science slipping into nonsense. No man has come into existence without a mother, and certainly none have shown signs of missing a rib. Moreover, in the light of the phenomenal revelations, *('Quantum Theory'- Qualitative Dynamics)*, concerning the emergent properties of *'Autopoiesis'*, showing that every cell in the living body knows how to make itself, but no part-such as a rib could be separated to complete this wondrous transformation of *'macromolecular arrangement & specific structure':* *'Life Unfolding: 'The Making of the Human Body': Jamie A. Davies).* All parts must be involved in this dialogue with, *'The Mind of our Creator'.* Whilst we have become aware that scientific thought appears to be converging with the Teaching of Jesus, it seems that *'Theological and Theosophical'* perceptions, relating to the role and status of women, may be lagging behind new scientific insight. Jesus condemned the heartless attitude towards women, promulgated by those who asked about their rights to divorce, showing that they regarded women as inferior and subservient to men. *('It was because you were so unteachable, but God created male and female: that is why a man must leave father and mother, and the 2 become one body':* *Mark 10.1-12).* They are made equal in the Sacrament of Marriage, and no one can put asunder what God has put together. Also, Jesus showed deep respect for all the women mentioned in the Gospels, making a special point of treating them with gentle understanding: e.g. In His conversation with the Samaritan lady at the well, and in the love he showed for his special friends, Martha & Mary and their brother. Furthermore, the disastrous consequences, caused by devaluing women and womanhood to an inferior and subservient status, are starkly evident in our modern world: e.g. The crisis of demographic imbalance in China, resulting from the practice of destroying baby girls and promoting one-child families. Following suppression of the natural maternal instinct, young women no longer express the desire to have children, causing an impending crisis of a shortage of human resources in the younger generation to support the needs of an ageing society.

Woman as a vessel of sin and temptation: Probably the most insidious calumny propagated against women in ancient Jewish scriptures, as recorded in the book of Genesis, this fallacious concept continues to act as a corrosive influence, with the effect of diminishing respect for the Holy Sacrament of Marriage, as ordained by *'Word of God'.* The destructive effects of this psychological weapon of hate have reduced the natural loving bond, which unites man and woman *'as one body',* to the farcical status of a casual agreement which can be broken at will. The knock-on effects of this hateful message are plain to see in our modern world: *('in 2016, the number of religious marriages in the UK. reached the lowest level ever recorded, (60,069), falling by 4.1% from 2015, and by 48% from 2 decades ago': Office of Nat. Statistics): (In 1986 40,000 Irish adults suffered a broken marriage. By 2011 this had soared to 250,000, affecting 300,000 children, bearing the traumatic effects resulting from a broken family: Irish Catholic News.).* In the Middle East this message of hate has left millions of orphaned children without parents, to fend for themselves. *(Gk. misogyny= hatred of women).*

Although these self-arranged 'civil' marriages may now be considered to be valid, because *'Marriage'* is the only sacrament that a <u>man and a woman can confer upon themselves,</u> they lack the understanding of empowerment which comes as a, *ray of 'Grace', in answer to an invitation to our Creator',* in dialogue with *'The Mind of God',* to bless the union and join in our human celebrations. At school, 85 years ago, we were told that such 'civil', arrangements could not be considered as valid marriages, and therefore were subject to a decree of annulment. It appears that in the course of time, *'theosophical judgement',* has softened with a more merciful sense of understanding human conditions. However, nothing was said to mitigate the misogynistic outlook towards women, advocated by St. Paul, arguably among the most zealous of the Apostles. *('For those who are not married it is a good thing for them to stay as they are, like me: but if they cannot control their sexual urges, they should get married, since it is better to be married than to be tortured; the man who sees that his daughter is married has done a good thing, but the man who keeps his daughter unmarried has done something even better':* Letter to the Corinthians:7:'Marriage & Virginity). Rather a grim and acerbic view of the world and personal relationships! We do not know why St. Paul came to view the natural relationship, formed by the bond of love between a man and woman with a jaundiced eye. We do know that he was one of the most murderous persecutors of those who became followers of Christ, until his conversion on the road to Damascus, to become one of the most ardent of the apostles.

(Gk. apostolus=messenger, from apostellein= to send forth).

Perhaps, this tendency to judge harshly, was the *'thorn in his flesh',* which he referred to as a penance to atone for this flaw in his psyche. He also makes it clear that the view of woman and marriage he was advocating, was his own, and not expressed to him by Jesus Christ. For proof of how Jesus viewed the God-given vocation of married life, we may imagine being amongst the guests invited to the, *'Marriage Feast at Cana': ('There were 6 stone water jars standing there, meant for the ablutions that are customary among the Jews... each could hold 20 or 30 gallons...when the mother of Jesus said to him, 'They have no wine', Jesus said to the servants, 'Fill them with water', and they filled them to the brim. 'Draw some out now', he told them, 'and take it to the steward': the steward tasted the water, and it had turned to wine': John : 2. 1- 12).* 180 gallons of the finest reserve wine? What a merry feast that must have been! Just imagine you had been invited, as one of the guests. How astonished, you would have been, to be tasting the most exquisitely palatable wine you had ever tasted? And in such abundance! This was the first miracle, enacted by Jesus, which demonstrated the divine status of the marriage union between a man and woman, as ordained by,' *The Word and Will of God'.*

This sends a message, which the younger generation- who are turning away from *'religious marriage',* would be well advised to take to heart, for their own good. When *we* invite the Creator, who brought us into being, to be with us and bless us in all we do, during the important events in 'Life Processes', (Marriage, Baptism, etc.), the response as a power of 'Grace', we may receive will be as astonishing as it was for those at the Wedding Feast at Cana. However, we must keep in mind that world events during the last century have created a fog of confusion in the mentation of young minds, quite puzzling to resolve. Recalling the *'Spiritual Retreats',* which took place regularly in my boy's school 85 years ago, with the intention of regaling us with an enlightened spiritual outlook, I still ponder the issues involved. Instead, they had the opposite effect upon young boys, in the confused state of early pubescence, because of the heavy emphasis on, *'sins of the flesh',* coupled with the caveat, *'to beware of dwelling on impure thoughts concerning the opposite sex',* with the implication that these were mortal sins, which we were required to purge by going regularly to, *'Confession'.* Inevitably, minds conditioned in this way become inclined to view the sense of attraction to the opposite sex, as sinful temptation, rather than a holy sacramental endowment, *as expressed in the religious Sacrament of Marriage.* Fortunately for me, my two sisters, 10 years younger- appearing to me like 2 little angels- provided a corrective influence, making me aware that*, valid knowledge may be transmitted as misleading information.*

In this way, perception of the world may be coloured, to appear as a place of conflict and hardship, causing many to lose faith in the concept of religiosity, as a *'binding force'* of brotherly love and holiness. *('You must not let sin reign in your mortal bodies or command your obedience to bodily passions, why you must not let any part of your body turn into an unholy weapon fighting on the side of sin; you should instead offer yourself to God, and consider yourselves dead men brought back to life: you should make every part of your body into a weapon fighting on the side of God; and then sin will no longer dominate your life since you are living in Grace'): St.Paul: 'Holiness, not sin to be the master': Romans 6. 12-14).*

In this respect, it is not easy for those living in poverty, striving just to survive and provide the bare necessities to keep their children alive, to remain so submissive. Jesus Christ demonstrated a more sympathetic attitude towards sinners; forgiving rather than condemning; healing rather than rubbing salt in the wounds; especially respectful of the human conditions which beset women's *'life processes.* In his conversation with the Samaritan woman at the well, instead of expressing disapproval of her way of life, and rather saucy manner, He commanded her attention and respect by gently chiding her boldness in a good-humoured way. Thus, He not only opened her mind to, *'know herself'*-probably for the first time in her life, but also infused an empowering spirituality of *'Faith' into her cognitive outlook'. ('Come and see the man who has told me everything I ever did. I wonder if He is the Christ. This brought people out of the town, and they started walking towards him'): John 4. 1-42). ('For God sent his Son into the world, not to condemn the world, but so that through him the world might be saved'): John 3. 1-21).* The significance of this profoundly important message has been obscured by contentious *religious opinions and arguments.*

Secular World: Religious Arguments Divide: Faith Unites: When I first went with my family to visit my grandparents in Ireland, over 90 years ago, it was like going into a different world that had existed long ago. Walking with my Granny to attend Mass, in the ancient Dominican Church in Kilkenny, I noticed that everyone seemed to be united, with strong devotion to their **Faith.** The Black Abbey church was full of people, waiting for Confession, which was available every Saturday, for several hours. Frequently, sitting beside the stove in my Granny's kitchen, there would be a knock on the door, and a vagrant would appear, in a tattered worn coat, held together by string: *"Would there be a cuppa tea, Missus'? 'Come in Paddy, sure you're always welcome'.* As if it were quite normal, my Granny would fill a bowl with delicious stew, from a pot that seemed to be always bubbling on the stove.

She told me later that he had suffered greatly, under the brutal treatment meted out by the vicious *'Black and Tans',* the military brigade formed by releasing criminals held in British jails- including many condemned to life imprisonment for the most heinous crimes, granted amnesty for the purpose of subduing the remaining population, already decimated by starvation caused by the potato blight and subsequent famine, into abject submission. He had been a prosperous farmer but had lost everything-including his home and family. He was reduced to a life of extreme poverty, sleeping in haystacks, and having to beg for food. Nevertheless, he had retained his faith, and still found solace in attending Mass, providing he stayed at the back of the church. This image of my grandmother, as an elegant gentle lady, remains indelibly impressed in my memory, the very embodiment of Christ's words, *('Anyone who does not welcome the Kingdom of God like **a little child will never enter it'**): Luke 18.13-17).*

That is how the spiritual quality of *faith,* once evoked may

may be sustained in the cognitive outlook of the **human mind in dialogue with the Mind of God,** capable of acting like an autocatalytic reaction, with the profectitious effect of replacing the mood of baleful pessimism affecting the younger generation throughout the world, with an outlook of hopeful optimism. Reaching out with loving care to others in need, as my grandmother showed me, calls forth the same response.

('The wisdom that comes from above is essentially something pure; it also makes for peace, and is kindly and considerate; it is full of compassion, and shows itself by doing good; nor is there any trace of partiality or hypocrisy in it. Peacemakers sow the seeds which will bear fruit in holiness'....'Pure unspoilt religion, in the eyes of God our father is this, coming to the help of orphans and widows when they need it, and keeping oneself uncontaminated by the world': James: 3.17-18. & 1.27).

First Law of Karma: **'As you sow, so you shall reap; this is the reaction of cause and effect. Whatever you put out in the Universe will come back to us. if you want happiness, peace, love, friendship-then be happy, peaceful, loving and a true friend'. The 2nd Commandment is like the 1st: *'You must love your neighbour as yourself': Mat.22.34-40.***

Convergence of Science with *'The Mind of God:* Advances in Science, such as the revelations of *Qualitative Electrodynamics, (Quantum Theory),* show that nothing in the Universe can now be considered as solid material, but that all we perceive, observe and believe, turns out to be the emergence of abstract probabilities, ('atoms'), as properties of molecular systems, such as the, unique, *'macro-molecular structure', which brings forth each living thing'.* Only the **human mind has been endowed with the cognitive capacity to comprehend the significance, *('semantic knowledge'), and the message transmitted, ('semiotic understanding'): 'that what we perceive as scientific observation could have no meaning without the ability of the mind to know itself, as a link in the chain of interactions observed and perceived.* The observing human mind is a necessary interacting process in the systemic dynamics with which it is engaged', a higher spiritual quality of guidance and control, denoted as, *'Freewill', inspired by 'Love & Truth'*.

In this sense, we are all '**religious**', (bound together), even those who profess not to be so. *('It comes from the father of all light, with whom there is no shadow of a change: By His own choice he made us His children by the message of truth, so that we should be a source of first fruits of all that He has created. Accept and submit to the word that has been planted in you and can save your soul. But you must do what the Word tells you, and not just listen to it and deceive yourself': Letter of St. James).*

It makes good sense to take care of our home and each other, for the good and well-being of all. When next I visited my Granny, aged 80 in 1946-at the end of WW2, she had retained her sense of childlike innocence and faith, but there was a marked change in the social mores and religious outlook of the people. Fewer were going to Mass, and the queues for Confession were no longer to be seen. Men returning from military service, embittered by traumatic experiences, and weakened in faith as a result, had lost the will to continue in religious practice. Young men and women were emigrating in droves, to escape the grinding impoverishment resulting from the debilitating effects of genocidal rule, and the ensuing famine. The population, numbering 9 million in the 16th century, had been reduced to about 2.5 million. Drawn by the insatiable demand for cheap labour in Britain, the Irish diaspora provided the vigorous young workers needed to supply the infrastructural requirements of the Industrial Revolution: e.g., The Nursing Profession; Construction of roads, railways & housing, etc. A redistribution of the population was taking place in England, as rural agricultural workers were being replaced by machines, and the demand for industrial labour in the factories springing up in towns and cities, grew rapidly: (*'In England's dark satanic mills':* *'Jerusalem': Wm. Blake).* The offer of vastly higher wages created an irresistible push-pull effect, but the lethal effects upon health, as a consequence of living and working in heavily polluted and wearing conditions have been concealed, and downplayed ever since. Continuing to this day, softened by the deceptive euphemism, *'Globalization'.* Instead of being a development of international communications as it has been represented, it may be seen as *'trawling the world for cheap labour',* with little regard for the workers involved. Thus, those attracted to work as coalminers have been exposed to *'silicosis', fibrosis of the lungs',* caused by inhaling dust particles; workers handling asbestos have contracted *'asbestosis & mesothelioma',* causing irreversible damage to the tissue around the lungs and stomach, caused by imbibing fibres of asbestos, formerly used in the building industry for insulation and as a fire retardant. **'Globalization',** must be recognized and countered, as a pernicious divisive influence, which has created the insidious, *'gap of inequality',* between those who have amassed inordinate wealth and the poorest workers. By taking advantage of advances in industrial and technological developments, workers have been cajoled to provide the labour which has made some entrepreneurs exorbitantly rich, by having to work in extremely dangerous and injurious environmental conditions, with the spurious assurance that they too will become well-off. Like a double-edged sword, this false ideology threatens to cascade into an unmanageable divisive force, causing rumbling discontent among workers throughout the world, who- no matter how long and hard they work, find it hard to make ends meet- (*'The just about managing workers'),* as well as souring relations between labour, and the highly remunerated managerial class who relentlessly call for, *'an increase in productivity',* from already hard-pressed workers. In this way, human resources become devalued as *'slave labour',* considered as expendable, justified by the spurious reasoning of *'eugenics & ethnic cleansing'*. Ultimately, the bond of **love** may be replaced by the hostility of **hate**. This corrosive influence may be seen as a perversion of the meaning and significance of the 2nd commandment, is gathering momentum in the world around us, acting as a malignant force of evil, and causing utter environmental destruction, as well as immense loss of life. *('Peter went up to him and said, "Lord, how often must I forgive my brother if he wrongs me? As often as 7 times?" Jesus answered, "Not seven, I tell you, but seventy-seven times"): Matt. 18. 21-22.*

It is easier to forgive, when the offence perceived is of a minor nature, such as a family dispute, or your neighbour's dog running around your garden, leaving a trail of havoc: not so easy, at a larger regional scale, especially following the destructive depredations of war, caused by a neighbouring country, involving grievous loss of life. When the cohesive influence of **brotherly love** becomes diminished, or blotted out by motives of **hatred and revenge,** the inevitable consequences are to be seen in the world around us, tarnished by **environmental destruction, and collapsing systems of organization:** e.g., in Syria at present, and in Ireland during the last 2 centuries.

Coronavirus: (Covid 19): Global Cases 1,039,1666: Deaths 55,092: *(World News: 3 April 2020).*

Environmental destruction paves the way for the onset of pathogens, such as, *'Bubonic Plague'*, *'Ebola'*, & *'Covid 19'*. For example, the ravages of the Mongol horde, led by the murderous psychopath Genghis Khan, left a wasteland littered with an estimated 20 million dead, about a third of the Eurasian population at that time. The ensuing *'Black Death'* plague, a rat borne haemorrhagic disease, wiped out a third of the population in Britain during the 14th century. It cannot be considered just a coincidence, that the *'Covid 19'* disease has emerged after putrefaction of the Eurasian & Middle East environment on an even greater scale, by continuous warfare, with millions of desperate migrants trying to survive in encampments amid filthy unhygienic conditions, like the 'jungle camps' around Calais and Dunkirk, and the rubble strewn wasteland round Idlib in Syria, once a thriving agricultural region, now a graveyard. These are the conditions which favour a rapid proliferation of the rat population, and the onset of disease. No wonder the younger generation are becoming disillusioned to the point of losing faith in government, as well as the tenets of religious practice, becoming convinced *'that we live in a secular world'*, where no one can be trusted. Thus, ***'The Word of God'***, may be obscured, and ***essential knowledge*** subverted to ***misinformation.***

However, the voice of our **Creator** cannot be silenced because the human mind was formed and endowed to hear and understand the meaning of *'The Divine Words', with the spiritual capacity to communicate with the **'Mind of God'.*** Valued and understood as a*' Reality of Geographical Existence'*, equal in significance to *'Life Processes', and 'Environmental influences',* mankind would be well advised to pay heed to the message we are receiving in terms of the dystopian effects of reckless and uncaring human activities: e.g. *Crisis of climate change; Environmental degradation through the ravages of war; Devaluation of Life and human beings; Souring of social relationships; Onset of Pandemic Disease; & Collapse in systems of human organization.*

Christ's words, "Love one another as you love yourself"*,** may be heard ringing with even greater insistence, echoing the collective theosophical wisdom of old; ***'Whatever you do, however you think, will come back to you: you do it to yourself! The every same message which rocked scientific thinking to its foundations, just before the outbreak of the 2nd World War-*'**Qualitative Electrodynamics'**,(Quantum*

Theory). All that we think we know and perceive must now be understood as the emergence of the highest probability; not in a material sense, but as the mysterious properties of molecular systems, which have been shaped by what has gone before, including the aspirations-whether good or evil affecting human behaviour and activities: i.e. *'We shall reap what we have sown'.*

Therefore, it should be obvious that perception of ourselves, just as material entities living in a material world, *'where everything is up for the taking, and self comes first',* is an erroneous and destructive concept, which generates the corruptive sense of hostility and hatred we observe erupting in regional and national relationships. Thus, the ***meaning of Christ's words, can be obliterated.*** *'Love',* by contrast is a spiritual cognitive sense of awareness, which attracts and binds people together, a truly ***religious empowerment of the human mind,*** which generates the trusting optimistic outlook of ***'Faith'.*** *(L. religare= to bind, join together).*

Moreover, ***the significance of Christ's message has been obfuscated,*** by cultivating an antagonistic and pessimistic outlook towards other people, and what is happening in the world around us. ***('You men of little faith! But you must not set your heart on things to eat and things to drink; nor must you worry. It is the pagans of this world who set their hearts on all these things. Your Father well knows you need them, and these other things will be given you as well. There is no need to be afraid , little flock, for it has pleased your Father to give you the kingdom').*** *Luke :'Trust in Providence'; 12. 22-32).*

There is no hint of condemnation or wrathful judgement in these words, full of loving forgiveness. Why then are the younger generation falling easy prey to the nihilistic belief that our *Father Creator is o be feared as a vengeful God, the very reverse in meaning to Christ' s persistent message?*

Rather, **Our Father** should be revered and trusted, (*'Providence'*), and loved as the beneficent creator of Life, and all living things, as well as our benign world environment, just right to support all our needs. Jesus told us that in a few words, and on many occasions during his ministry on earth.

('Ask and it will be given to you. Is there a man among you who would hand his son a stone when he asked for bread. If you, then, who are evil, know how to give your children what is good, how much more will your Father in heaven give good things to those who ask him!': Matt. 7. 7-11). There can be no doubt that we are confronting symptoms of impending disorder in *'Life and World Systems'*, such as the *'Covid Pandemic'*, which threaten to cascade to an unmanageable scale. Therefore, it should be obvious to all, even the most avaricious and self-seeking, that it makes good geographical sense to look after our home-world, and to cultivate good relations with our neighbours. If we destroy them, we destroy ourselves. We never know when we shall need their help. Why then, are our younger generation so easily persuaded that, *'we live in a secular world, wherein it makes sense to put yourself first'?* Divisive concepts lead to a fearful outlook towards a world which appears to be a dangerous place, as well as a mistrustful view of our fellow human beings. Like a child watching a horror film, taking refuge behind the sofa when the *'alien cybermen'* appear, the human mind may be distorted, or fatally depressed by what may appear too alarming to bear. In this way, the higher cognitive processes of *'**spirituality of mind'**, discernment, aesthetic appreciation, & freewill,* are susceptible to disruption, and ultimately to impairment so extreme, that **'Life'** may be perceived as distasteful and not worthwhile: e.g., increasing cases of psychological disturbance, self-harming, and juvenile suicide.

Fig.5: 'Quantum Biological Connections': This presents a new, more comprehensive view of Life in our Earth-world, resulting from the combined research of scientists, relating to *'Complex Networks of Interaction'*, *'Fluid Dynamics'*, & *'Chaos Theory'*, defining a new *'Ecological Science of Applied Geography'*, where *'The Human Mind as a System of Control'*, is of focal importance

Fig.6: 'My forgiving Irish Grandparents': The sense of *'Ecological Care'*, together with a tolerant forgiving outlook, make for peace and harmony in *'Systems of Human Organization'*, stemming from *'The Cognitive Process of Awareness of the Bond of Love'* in social relations.

(Gk. empathea from pathos=feeling, the cognitive ability to care for others

khatharsis from khatharen =to cleanse, from khatharos = pure, the state of healing, immuno-logical reponse

phainomenom from phanein= to show, demonstrate something, reveal the meaning

ekkentros= unusual, unexpected, whimsical, out of the ordinary

phantasia from phantazein=to make visible, appearance, cognitive process of imagination

kosmos= order, omament, world, harmony, concept of Universal Order, ordered system of ideas

apatheia from pathein= to suffer, without feeling, indifferece to grief or plight of others (penthos=grief).

kuklos=recurring, cyclical: daemonikos from daemon=evil spirit, deviish, wickedness

symptoma from piptein=to fall, deteriorate,=signs of degeneration: holos=whole, entire

parallel from allellos=one another, similar, side by side, comparable

antithesis from tithenai=set, place, +anti =the opposite of what has been set down, opposite of what is known as ordered or expected= disorder; khaos= vast chasm, chaos, extreme disorder

Figure 5: Quantum Dynamics : Quantum Biological Connections: Patterns of organization and regulation: A dance of interacting parts.

Enzymes : The catalysts which speed up interactions: autocatalysis-they can make themselves:
Enzymes make and unmake every single biomolecule inside every living cell that lives, or has ever lived. Note:- At the nanometre scale (millionths of a millimetre, in this space **o** you would see the whole of USA. This would let you see the human cell as packed with water molecules &metal ions. An ion is an atom or molecule with a non-zero electrical charge-its total number of electrons is not equal to its total number of protons.

Amino-acid molecules slide up and down collagenase strings, and can alter peptide chains by snipping chemical bonds, capturing electron energy or protons, creating a proton gradient (ie., like a flow downhill from outside to inside of the mitochondria.
Therefore, the key events of respiration have more to do with an orderly transition of electrons through a relay of enzymes inside our cells.   The puzzle is, how do these enzymes shift the electrons so quickly and efficiently across big molecular gaps. *Devault and Chance(1966),* showed that the rate of electron hopping did not drop at low temperatures.

Tunnelling: Atoms are not things, but probabilities which come into existence in molecular systems: they can overcome energy barriers: An atom, once connected with another, remains in communication, even if separated to the end of the universe: nothing is solid in the universe:
Quantum tunnelling allows particles to pass through impenetrable barriers- like sound waves through walls, the same process which allows positively charged H nuclei to fuse together when hydrogen converts to helium in the Sun.   Newtonian mechanics requires an extra kick of energy to climb the hill of a chemical barrier, but quantum tunnelling allows the electron to flow through as a wave: Energy can do this.   Also, the lighter the particle, the easier for it to tunnel.   That also explains how radioactive decay takes place.

Coherence: molecular systems appear to obey a principle of ordered regulation, even falling into single file (e.g. when having to pass through microtubules in the brain):
Quantum tunnelling depends on the spread-out wave-like nature of particles, but with many particles, all the constituents have to march in step, for example:- in phase change, such as magnetization, crystallization, etc., all particles must follow the same orientation, to maintain order.

Decoherence: Explains how patterns of regularity can be linked with non-linear feed back loops of interconnection (patterns of aperiodicity linked with patterns of regularity): Theories of Complexity and Universality (A fundamental Principle of Order common to all systems;
This is the process by which all the many quantum waves get out of step: the jostling and jiggling turbulence of atoms and molecules.   Therefore, big objects consisting of trillions of atoms should not be able to Quantum Tunnel.   Living cells are big objects and should not be able to tunnel, especially in hot wet conditions in which living cells live, but they do, even protons (2,000 times larger than an electron).   Therefore, the interior of an enzyme is different: it is engaged in a choreographical dance as if designed from the beginning to achieve a result. Note that every cell has the information (code) to produce some 2,000 enzymes to act as chemical/energy messengers to achieve a particular purpose in the functioning of systems in order to support Life.

(Martin Rees:'Just 6 Numbers': The Universe appears to be 'fine tuned' from the beginning to support Life).

(Richard Feynman: 'Anyone who claims to understand Quantum Electrodynamics certainly doesn't)

Fig. 6: My forgiving Irish Grandparents: Kilkenny 1953

'*The Kilkenny People: 1957':* CELEBRATES 90th BIRTHDAY: 'It is quite some
time since John Denieffe, who lives at 4, Blackmill Street, first heard the piercing ring
of an anvil. For this grand old man, with a light-hearted smile and quick sense of humour
was a blacksmith all his life, as were his father, grandfather and great-grandfather before
him. His wife Margaret is 81, and they had 14 children, 7 of whom are dead.. Many fine
examples of his work are to be seen in Kilkenny, such as the railings round St. Kieran's
cemetery, as well as the magnificent entrance gates, and the iron work at the statue of
Our Lady, at The Black Abbey Dominican Church.'

Cybernetics:
Science of Control and Guidance

Pleasant and loving thoughts unite: Frightening and disturbing thoughts divide:

Bereft of *'Cybernetic' systems of guidance and control'*, the human mind is reduced to the state of a ship without a rudder, tossed and tugged in all directions in a turbulent ocean of misinformation, disinformation, sophistry of specious reasoning, deceptive semiotic arguments, and misinterpretation of language *(semantic reasoning)*.

(Gk: sema=sign, semantikos= meaning of language: semiotikos= significance, message).

In this context, the ***internet,*** which has enabled the development of a global network, providing world-wide communication facilities, appears to have become *'the devil's playground'*, for leading young minds astray. Acting like a *'cognitive virus'*, armed for the special purpose of infiltrating the *'Cognitive Processes'*, the faculties of *'understanding and comprehension'* are being corrupted, with the effect of confusing and obscuring communication, concerning ***'The Words of God'***. Just as *'Covid 19'*, appears to have the ability to mutate into an even more aggressive pathogenetic disease, this could be regarded as an even more threatening ***'hyperparasitic'*** mental affliction, capable of deadening all sense of empathetic redeeming ***faith,*** amounting to a blasphemous negation of the empowerment of ***The Holy Spirit.*** *(Gk. blasphemos = evil speaking).*

Jesus hardly ever spoke words of condemnation, when confronting sinful human frailty, but in respect of sins against ***The Holy Spirit,*** He was adamant: ***('I tell you, if anyone openly declares himself for me in the presence of men, the Son of Man will declare for Him in the sight of God's*** *angels. But the man who disowns me in the presence of men will be disowned in the presence of God's angels. Everyone who says a word against the Son of Man will be forgiven, but he who blasphemes against the Holy Spirit will not be forgiven. Do not worry about how to defend yourselves or what to say, because when the time comes, the Holy Spirit will teach you what you must say'): Luke 12. 8-12).*

Moreover, parallel developments in the spurious science of *'AI': Artificial Intelligence,* appear to be aiming at the same objective: viz., replacement of the human mind with electronic devices, for the purpose of making decisions, based on the fallacious and misleading premise, *'that these can be programmed to carry out routine tasks more efficiently'*, as well as making life easier and more enjoyable, by relieving us of the need to remember minute details. The truth is, that there can be no such thing as *'artificial'* intelligence. *Intelligence*, the cognitive faculty of understanding, and *intellect,* the complex process of cognitive immunological guidance and protection endowed in the human mind, must be regarded as precious as *Life* itself; seen and recognized, as the working of *the Holy Spirit, in the consciousness* of every human being: *i.e., Internal valid knowledge,* **Faith.** There Is an even more insidious viral menace, looming on the horizon, in the seductive form of an enjoyable game and pastime. In fact, it should be recognized as another aspect of the same dark force, probing the way to infiltrate and dominate the human *'psyche',* and ultimately to predominate in the **Universe.***(Gk.psukhein=tobreathe;psukhe=breath,soul,spirit, Life}.*

Vaunted by the *'Media',* as *'Virtual World experience',* it has branched out into the more destructive and enticing viral influence of *addictive gambling:* e.g., *Fortnite World Cup: 29 July 2019: ('16 year old boy wins $3 million in online 'Shooter Game').* Gambling has become the scourge of our time, proliferating worldwide as an obscenely corrupting influence, especially affecting the younger generation and the poorest, with the lure of becoming rich. Instead of becoming rich, countless numbers, including children playing gaming machines in amusement arcades, have been burdened with debt, leading to a life of misery, as well as vulnerable to abuse and enslavement. All sports have been drawn into this moneymaking fraudulent network of complicity, resulting in further dystopian economic effects: e.g., a scarcely literate footballer is valued as worth more in a week, than 40 State Registered Nurses are paid in a year.

Christ's Words should be heard rising to a crescendo: *('Obstacles are sure to come, but alas for the one who provides them! It would be better for him to be thrown into the sea with a millstone put round his neck than that he should lead astray a single one of these little ones. Watch yourselves'): Luke 7. 1-3). ('Yes, I tell you again, it is easier for a camel to pass through the eye of a needle than for the rich to enter the kingdom of heaven': Matt. 19. 24-26). ('What doth it profit a man if he should gain the whole world* <u>*and suffer the loss of his own soul?)*</u>.

Christ's challenge to the world: Christ's promise to those who retain their faith & trust in Him: *('But I say this to you who are listening: Love your enemies, do good to those who hate you, bless those who curse you, pray for those who treat you badly. To the man who slaps you on one cheek, present the other cheek too. If you love those who love you, what thanks can you expect? Even sinners love those who love them. You will have great reward, and you will be sons of the Most High, for* <u>*He Himself is kind to the ungrateful and the wicked'):*</u> *Luke 6.27-35).*

Is it really possible for those who have been subjected to extremely barbaric suffering, such as the traumatic cruelty experienced by millions, under hateful Nazi rule, in World War 2? **Yes, it is!**. For example, in 1941, Fr. Maximillian

Kolbe, a Franciscan Friar imprisoned in Auschwitz Death Camp for disobeying a German order, gave his life willingly to save another man, who had been selected for execution as a reprisal after a prisoner had escaped. He was canonized by Pope John Paul 2, as a martyr of charity, on 10 October 1982. Also, I recall that my Granny displayed similar generosity of spirit. in her night prayers-after The Rosary, she would include a prayer for the *'souls of those most in need of God's mercy'*, especially those responsible for the dreadful atrocities she had witnessed, resulting in the death of her own children and countless other people she had known, during the Famine years. **Loving thoughts engender the spiritual qualities of *'faith & ecological care'*, needed to restore order to our disturbed world, and to repair the ruinous effects of environmental degradation.**

Dichotomy of a divided world: Christ's Prayer:-'Ut Omnes Unum Sint:- All be One' Faced with signs of mounting disquiet and increasing disorder in the world, it is not surprising that leading scientific analysts have taken a pessimistic view of the probable outcome: e.g. Lord Martin Rees, Astronomer Royal: (*'Is the 21st Century our last?'*). However, there is another more hopeful view, that addresses the *'phenomenon of consciousness'*, imbued with the redeeming power to counter problems of turbulent disorder.

The developing science of **Cybernetics** questions the corrosive philosophical outlook of *'materialism'*, the belief that *'consciousness and will'* are governed wholly by material influences, with an all-consuming desire for material possessions and physical comfort. In that way of life spiritual values have become diminished, supplanted by material interests. Thus, the current obsession with *'Artificial Intelligence'*, promoted by the mass media, may be regarded as a spurious, misleading ideology of communication, rather than a science of valid information.

('The arguments I am presenting point to several places where our present-day pictures fall profoundly short of providing us with a scientific understanding of human mentality—there should indeed be a scientific path to the understanding of mental phenomena, and that this should start with a deeper appreciation of the nature of physical reality itself. I feel it is important that any dedicated reader, wishing to comprehend how such a strange phenomenon as the mind can be understood in terms of a material physical world, should gain some significant appreciation of how strange indeed are the rules that must <u>actually</u> govern that <u>material</u> of our physical world'.) Sir Roger Penrose: Pioneer Mathematician of 'Quantum Theory': 'Shadows of the Mind': Preface. **(19).** This brilliant study makes essential reading, as an enriched flow of epistemological knowledge, uniting true religious, and scientific reasoning to speak with one voice.

Thus, catastrophic events, such as 'Covid19 Pandemic' and 'Climate Change', may be pessimistically construed as harbingers of crumbling world disorder. A more constructive outlook is to view the wide swings we are perceiving in world systems shaping our daily lives, as chaotic fluctuations which could be managed to restore order, according to a *'Universal Principle of Order'*, **inherent in all systems.** Known as, 'The Butterfly Effect', computer models of such systemic chaotic swings, appear as an octal pattern, forming loops around 2 points, dubbed 'strange attractors', with the effect of pulling the system back to order.

Episodes of Disorder: Cataclysmic events, such as the outbreak of 'Covid 19', tend to evoke 2 reactions: the first, is to give way to panic, driven by anxiety and self-concern. A more considerate and reflective response is to realize 'that we are all in it together', so that it makes common sense to unite, in devising effective strategies to counter and mitigate the probability of more disastrous effects which might arise. In that way, we may appreciate the power of, *enlightenment in*

'Logical thinking and Spirituality of understanding', latent and available to us for our own good, if only we have the *faith to know and make use of it.*

('The godless call with deed and word for Death. Come then, let us enjoy what good things there are, use this creation with the zest of youth. Let our strength be the yardstick of virtue, since weakness argues its own futility. Care for discipline in your life comes from wisdom: watch for wisdom in your life, and you will have no trouble, you will find <u>her</u> sitting at your gates. Even to think about her is understanding fully grown; be on the alert for her and anxiety will quickly leave you. <u>She herself walks about looking for those who are worthy of her</u>': The Book of Wisdom.*

This is the great lesson to be learned in the course of our lives, the challenge presented in ,*'The Word of God', and reiterated in the parables of Jesus Christ. What has gone before, should serve to point the way forward!* But minds wounded by traumatic experiences tend to develop a sense of immune amnesia, to forget unbearable memories.

For example, the world situation was probably worse in the 1930's, in the aftermath of the 1st World War, than it appears to be now. After the loss of millions of young men, leaving countless families across the world shattered with grief for the loss of the breadwinner, as well as in a state of destitution, there followed *'The Spanish Flu Pandemic',* which has been reported to have wiped out about another 100 million people. The Great Depression ensued 10 years later, leading to a severe economic collapse in systems of banking and economic organization throughout the world, lasted until the outbreak of the 2nd World War in 1939. The high probability is that it was caused by printing cheap money, to ward off the inflationary effects triggered by the enormous costs of total war.

Then followed a winter of discontent, as the working wage was cut to the breadline. Workers, such as the Jarrow coalminers, who had toiled so hard to meet the demands of war, unable to earn enough to feed their families, rose in protest in a *'hunger march'* to London. Since then, *'Economics'*, appears to have become a spurious science of monetary & financial organization, focused on apportioning debt, rather than wise management of financial systems. Peering through the mists of history, it would seem that this scenario of cascading turbulence, from hostility, hatred and war, to the dystopia of economic ruin, leading to smouldering resentment amongst the impoverished working population, has been repeated at shorter and shorter intervals, time and time again.

– The Antonine Plague, 165-180 AD, brought back to Rome by returning troops, caused the death of 5 million people, and killed 10% of the Roman population. Probably measles or small-pox, it devastated the Roman Army, and led to a long period of unrest and political instability throughout Eurasia, culminating in the disintegration of the Roman Empire in the 5th Century AD, as well as disruption in the Catholic assembly of Christians, which had spread, united and in harmony across the Mediterranean and Middle Eastern region, until the Nestorian Schism, 431-544 AD. This involved a difference in opinion between Nestorius (Patriarch of Constantinople), and Cyril (Patriarch of Alexandria), concerning the human and divine natures of Christ. Condemned by the 1st Council of Ephesus in 431, and the Council of Chalcedon in 451, this dispute flared into a breach between the State Church of the Roman Empire, established by Constantine, and the Eastern Sassanid Persian Church, which persists as a divisive barrier between the Orthodox and Catholic Assemblies to this day. The question arises, why were the Nestorian brotherhood treated so harshly, that they were forced into exile, out along the Silk Road, *('The Golden Road to Samarkand'),* where they were made welcome by Eastern Philosophers in communities of mixed race, living in peace with each other for centuries, such as that established at Turpan Oasis, in Uzbekistan.

-- The rift between Eastern and Western Christian faithful assemblies was further exacerbated, and relationships more embittered, when Pope Urban 2 ordered the 1st Crusade, (27 Nov. 1095), in response to the call for help by the Byzantine Emperor Alexius 1, (1081-1118), against the depredations of marauding Muslim Seljuks. With the words, *'Deus vult' (God wills),* this command, as if expressing *'God's Words'*, sanctioned the onslaught of an unruly pillaging mob with little regard for hallowed places, in direct contravention of *Christ's Words*, *('Love your enemies as if they were your brothers').* The Holy Land has been desecrated by hateful episodes of conflict and contention ever since. In 1204, the ancient Greek city of Stamboul (Stampol), founded in 660 BC, and chosen as the capital of the Eastern Roman Empire by Constantine (Constantinople), was sacked and plundered in the 4th Crusade.

-- In 1260, the ravages of the murderous Mongolic horde left a train of hideous bloodshed and devastation, with environmental conditions conducive to form a breeding ground for the *'Bubonic Plague Pandemic'*, which eventually spread across the world during the 14th Century, including Europe, Asia and Africa. Caused by the bacillus, *'yersinia pestis'*, which is found in animals throughout the world, and spread by fleas living particularly on rodents- such as rats, mice and squirrels, it appeared in 3 forms (*'Bubonic, Septicaemic & Pneumonic'),* and has caused a devastating loss of life. Estimated to have killed 50 million people in the 14th Century, amounting to a loss of over 50% of the population in Eurasia and Britain, it also led to a long period of severe economic decline, culminating in rebellious uprising of the working population, who had suffered the most, as a result of this deadly pandemic plague. The faint echoes of the anguish and fear which must have clouded the minds, even of children, at that time, can be heard still in the play of children to this day: *'Ring a ring of roses, a pocket full of posies, atishoo, atishoo, we all fall down'*, mimicking the symptoms observed in those collapsing around them.

In 1453, the Byzantine Empire, the last remnant of the Catholic assembly of the Roman State, fatally weakened, came to an end under the domination of Sultan Mehmed of the Ottomans.

– The Bubonic Plague emerged again in 1665 in Britain, brought in by trading ships infested with the Eastern black rat, and spread rapidly from the plague epicentre in London. The country narrowly missed a further devastating loss of life, when the Great Fire of London broke out in a bakery located in Pudding Lane, in 1666, destroying the rodent nesting places, together with most of the wooden buildings, as well as the old Cathedral of St. Pauls and Old London Bridge.

-- Cholera, a bacterial gastrointestinal infection, caused by conditions of poor sanitation, and ingesting contaminated food and water, is a lurking and highly infectious disease. There have been 3 pandemic outbreaks during the last 2 centuries: 1817-1824 in India; 1826-1837 in the Middle East & Europe; 1837-1863 spreading to Russia where 1 million died, in a second peak and apparently more virulent form. This also emerged in London, causing 10,000 deaths. At the same time, the stench from excrement flowing into the Thames became so unbearable that parliamentary proceedings had to be suspended. London's entire sewage system had to be overhauled and replanned, by Sir Joseph Bazalgette. However, his plan was not fully completed, to meet the future needs of the developing underground sewage system, on the grounds of cost, until 1870, giving time for a 2nd Cholera peak to develop in 1860, at the cost of more lives.

-- Poliomyelitis, *(Gk. polios=grey + muelos=marrow: decay of bone marrow),* defined as an infectious viral disease affecting the central nervous system, which may cause lymphatic meningitis, or permanent – and frequently fatal paralysis, particularly of the lower limbs. Although thought to have dated from prehistory, it was identified as an endemic disease in the 1900's, *(Gk. demos=people; affecting people),* and developed into a major epidemic, which spread throughout Europe and Britain. It developed rapidly in the decade following the destructive ravages of war, reaching a peak between 1940-1950, and killing at least ½ million people each year, and disabling countless children. *(Gk. Epidemic from demos=people, prevalence of disease in a population).*

-- Tuberculosis, known as, *'phthisis',* and The White Plague, is the leading infectious cause of death in the world. Recognized from ancient times as a lethal pandemic, it kills a person every 20 seconds, and over 1.5 million each year. *(Gk.phthinein=waste away: "Epidemics" by Hippocrates).*

This insidious wasting disease is caused by the Mycobacterium bacillus, *(Gk. muces=fungus),* which affects the lungs, and other bodily parts: it can cause impotence. 70% to 90% of people infected by it show no symptoms, as with the *'Covid19'* virus, and appear to be able to fight it off, but it kills about 50% of those affected, and could be considered a more dangerous disease. It seems to be a long-lasting bacterium, capable of existing in xerophytic conditions: It has been detected, living between the pages of old books, in libraries of old houses, unused for up to 100 years. Some epidemiologists have questioned whether it may have developed, by *2 bacterial organisms becoming symbiotically united.* It merged as an epidemic in the 18h and 19th centuries, mostly in Europe and N. America. In 1870, it killed 20% of the entire population in N. America, and a similar number in Britain. Three stages of infective response have been noted:-
1) 70% to 90% have immune systems capable of fighting it

off; 2) Secondary or Reactive Stage: where immune systems have been weakened by mental, physical or emotional stress, *'Latent TB'*, can reactivate rapidly into active TB; 10 % of those diagnosed as latent TB, progress to active: 3) Active TB becomes highly infectious, passing through the blood stream, with fatal results: In 1913 it killed 117,000 in Britain: 1923, in England & Wales 32,000 died of TB of the lungs, and 8,000 from TB infection of other metabolic systems.

Biological Warfare: During the 2nd World War, a special unit was formed in the Japanese Army, Unit 731, to experiment and develop bacteriological strains that could be used as weapons of war, such as pneumonic-anthrax, possibly a combination of TB and the cattle disease, Anthrax: *('Mycobacterium + Bacillus Anthracis')*. Prisoners of war were used as guinea pigs, and the numbers killed by germ warfare in the Far East, consisting mostly of poorly equipped Chinese troops and civilians, has been estimated to be over 580,000: *(2002: International Symposium on Bacteriological Warfare)*. Lt. Gen. Shiro Ishii, the army doctor who carried out these atrocities, received immunity in exchange for the reports and data referring to these experiments, in 1948. Could it be just a coincidence that these biological weapons were used in the Battle of Wuhan, where a biological research laboratory still exists, and where the *' Covid 19 Virus'* first emerged? TB should therefore be considered as far more deadly and persistent than *'Covid 19'*, but one seldom hears it mentioned at Government level, as a sinister, lurking and ever-present pathogen, capable of obliterating a whole nation in one strike, by following the path of decay opened by other destructive wasting diseases, such as *'Covid 19'*. Why does no one seem to care?

Care on Earth: Generating Informed Concern: Statistical records show that wasting diseases inevitably follow, as the calamitous effects of destructive war and environmental degradation. We also know that we should learn from such traumatic experiences: '*Those who fail to learn the lessons of history are bound to repeat them!*'; & '*What you do to others, you do to yourself!*'. Yet, it would seem easier for the human mind to forget unbearable experiences, than to act with, **sensible logic of understanding,** let alone with the healing sense of, **spirituality of care.**

('Such wider vision requires even more complexity- that can value others not only in terms of helps and hurts, but also with concern for their health and integrity. This radically elaborates new levels of cultural information, and caring. Humans care about family, tribe, nation, careers, and ideational causes, such as biological science, French Literature, or the Christian faith. Ethics shapes caring. In due course, humans alone on the planet can take a transcending overview of the whole- <u>*and care for Life on Earth*</u>*...We need as full a story as we can get about caring on Earth. At length and in the end, we will seek the metaphysical and religious significance of this generation of caring':* Holmes Rolston, Distinguished Professor of Philosophy: Part3. Biology 11; pp 261-315: **'Information & Reality':** Editors: Davies, P., & Gregerson, N H.) **(52).**

The natural understandable human instinct is to soften memories of painful experiences, which are too stressful and harmful to keep in mind, by reconstructing them in '*language of information*', instead of worrisome '*valid knowledge*'. That is why men returning home after experiencing the abyssal horrors of war, rarely spoke about what they had been through, for many years. That is now recognized by neurologists, as a normal reaction to the effects of cognitive impairment.

'**Cognitive Processes**', including the faculties of *Language, Memory, Emotion, Perception, Freewill , Logic of Reasoning & Spirituality of Caring,* must be understood as intrinsically linked with, '**Life Processes**'. Consider the frenetic panic, in the rush to buy up all the stock in the Supermarket Stores, when *'Covid 19'* was declared to be a new dangerous pandemic disease, leaving the shelves bare of the necessities required by nurses and care workers who had laboured all day in hospitals and Care Homes, returning late on their way home in the evening. When minds are conditioned to behave selfishly, in this irrational manner, the spiritual sense of care throughout the community becomes depleted, with the effect of compounding problems calling for wise decisions we already face, only made worse by hasty foolish reactions. A few weeks later, rubbish bins became overloaded with the food, which had been snatched from nurses and health-carers at their time of need, as it began to decay, thereby creating a further environmental hazard. In this regard, it would be salutary to keep **Christ's words** in mind: **('Do not store up treasures for yourself on Earth but store up treasure for yourself in heaven; where your treasure is, there will your heart be also; So always treat others as you would like them to treat you; that is the meaning of the Law and the Prophets':** *Matthew 6. 19-21 & 7. 12).*

Other mindless reactions to the Covid 19 Pandemic: The probability of another Great Economic Depression became a matter of major concern, as the extent of the spreading disease was revealed by news broadcasts. Currencies fell in value, and panic selling caused a rapid decline in the value of shares on the Stock Market. But money is only worth the value we (Financial Authorities), choose to set on it. Why not therefore resolve this unnecessary tension by simply cancelling all debts, which have become so tortuously entangled, that no one understands what we are being forced to pay for, nor to whom the money goes? The mood of pessimism took hold, causing low-paid workers to clamour for a relaxation of the measures imposed to mitigate the rate of infection, regardless of the risk to themselves. The customary Christian observance of Lenten and Advent rituals, leading to the great Easter

Celebration of the Resurrection of Jesus Christ, seemed to take second place in news broadcasts, given just a passing mention, in contrast to the endless repetitive discussions devoted to, *'pointing the finger of blame'*, rather than to inspire a healing spirit of *faith & Hope.* We may learn much about ourselves by observing the reactions of others in times of stress, as the collective geographical knowledge and wisdom garnered from past experience tells us: (*'Know Thyself: All things in moderation!'*: *Thales (624-546 BC & Aristotle (384-322 BC).*

Invariably, hasty decisions aimed at self-preservation lead to a quagmire of despondency, ineptitude and inaction. Why should it prove to be so difficult for everyone to keep a clear head, and act with common sense, thinking with *'logic of mind'*, and *'spirituality of care for each other'*, knowing that all problems are best resolved in this way, for the good of all? That was the challenge transmitted in the *'The words of God'*, *spoken clearly by Jesus Christ: ('A good man draws good things from his store of goodness; a bad man draws bad things from his store of badness: by your words you will be acquitted, and by your words condemned'*: *Matt.12.35-37).*

Every human being is a geographer on a journey through life, intuitively in search of the best location to ensure a happy peaceful life for him/herself, friends, countrymen, in harmony with all people around us in the world. But it is not without risk. We all must inevitably gather a store of good and bad experiences, as well as memories of our own good/bad responses: *'Act in haste, repent at leisure'*: *'Look before you leap'.* **Christ's words echoed the wisdom of old: (*'When your eye is sound'*, your whole body too is filled with light; but when it is diseased your body too will be all darkness. If, therefore, your whole body is filled with light, and no trace of darkness, it will be light entirely, <u>as when the lamp shines on you with its rays'</u>:** *Luke 11. 34-36).*

These words show that it would be a fundamental travesty of the truth, to think of **God, the Creator of Life,** and all world systems, as wrathful and condemning. On the contrary, they must be understood as advisory words of *care and concern for the well-being of mankind, emanating from the absolute qualities of, 'Truth and Love', in the supernatural Holy Mind of God, as the 'Good Shepherd of His flock'. ('He guides me by paths of virtue for the sake of His name...though I pass through a gloomy valley, I fear no harm; beside me your rod and your staff are there, to hearten me': Psalm 23).* Gracious words of love and care inspire a spirit of optimism, faith and hope, when the path of Life leads through dark and stressful events, such as the *'Covid 19 Pandemic'.* This view of Life and the World, as one great ordered system points the logical way to deal with such problems, so as to resolve any crisis of management to arise.

Systems view of Pandemics: Cognitive Immune Response: From this point of view pandemic diseases may be perceived as *'Knowledge based systems',* charged with the task of clearing up decaying waste, which could be the cause of further harmful effects, to an already badly polluted and unhygienic environment. In this way, they would appear to be acting as supercharged immunological systems, rather like the *'Cytokine-induced killer cells (CIK cells)',* discovered in 1990 within the very complex human immune system, and found to have the ability to recognize infected or even malignant cells, in the absence of antibodies which are normally used to detect infection. This discovery has opened the way to a further development of *'induced memory cells (NK cells)',* acting like the natural *'CIK killer cells',* but with the enhanced ability to cause *'lysis or apoptosis',* only to specific tumours without destroying surrounding tissue. *(Gk. lusis= loosening-dissolving). (Gk. apoptosis= falling away, destruction of cells).*

Can there be any doubt that the human mind is being opened to comprehend what no one can possibly understand in entirety? (*'The incomprehensible thing about the Universe is that we can comprehend it': Einstein*). It seems that we are receiving, *'mental information including the key quality of semantics, which enables human beings to derive understanding of their world from sense data, and to communicate what we believe to be its meaning, to each other': 'Placed at the interface of the physical and cultural sciences, Biology plays a pivotal role in our understanding of the role of information in nature':* *Information & Nature of Reality: The Concept of Information in Biology': pp.157-186: Maynard Smith, J.: Geneticist and Evolutionist.* **(53)**

Could this be telling us that grim events, such as endemic disease, and major Pandemics, such as *'Covid 19'*, most probably emerge, according to an overall Principle of Universal Order, by which, *'all things created are directed to their appointed end!': Thomas Aquinas)?* Not as shallow scientific thinking is prone to assert, that evil experiences are incompatible with belief in, *'a beneficent creator'*, but as informational milestones, pointing the way through, *'vales of darkness', on the journey through life: (Psalm23).*

Understood as beneficial markers therefore, what appear to be calamitous events, such as *'Covid 19'*, could be construed as the emergence of a, *'Life-World immune response'*, for the purpose of eradicating and clearing up environmental putrefaction and decay, which left unattended, has the potential to cascade to even greater trauma, and destruction of life. Viewed in this way, *'vales of darkness and tribulation'*, could be considered as valuable lessons, it would be well to keep in mind.

('What Christians believe, and indeed other religious believers, today affirm as 'real' fails to generate any conviction among those who seek spiritual insight and who

continue regretfully as wistful agnostics in relation to the formulations of traditional religions...Many factors contribute to this state of affairs; one of these is that the traditional language in which much Christian theology, certainly in its Western form, has been and is cast is so saturated with terms that have a supernatural reference and colour that a culture accustomed to think in naturalistic terms conditioned by the power and prestige of the natural sciences, finds it increasingly difficult to attribute any plausibility to it': 'Information & Nature of Reality': pp.315-356: Arthur Peacocke: Biochemist & Theologian.

This brilliant insightful discussion illuminates how the Jewish followers of Jesus Christ reacted when they realised that *'something new had appeared in the world of immense significance for humanity',* but at the same time found it a challenge to reconcile the dichotomy presented by the *'dimension of divine transcendence they had attributed to God alone',* with the nature of *'a complete human being outstanding in every sense',* which they had encountered in Him. **'They had experienced an intensity of God's immanence in the world, different from anything else in their experience or tradition'.**

Put yourself in their shoes. The world around them was falling part yet again, under the fearsome onslaught of the Roman Legions, reduced to despair by the certainty of ruthless domination under Roman rule. They had experienced similar *'vales of darkness'* during the long periods of exile under Babylonian and Egyptian enslavement, as recorded in their Hebrew Scriptures, and knew only too well what to expect. Their only hope lay in the long-expected emergence of the, **'Messiah',** the one prophesied to liberate the Jewish Nation. *(Gk. Messias- from Aramaic: mesiha: from Hebrew: masha=to anoint: mashia= anointed). (Gk. prophetes=interpreter; from phenai=to speak: phetes=speaker).*

114

1. Therefore, when they heard of the miraculous events that were astonishing the growing throng following Jesus, such as the cure of the woman suffering from a continuous haemorrhage, that everyone was talking about, they hoped it would be *'the superman',* they had pictured as the **Messiah,** capable of repelling the might of Rome in one blow, *'making Israel great again'.(Matt. 9. 18-26).* The more scientific report, by a doctor trained to take a clinical view of **Life processes,** gives more information, adding valid knowledge which provides greater understanding, **(Cognitive enlightenment): ('A woman suffering from a haemorrhage for 12 years which no one had been able to cure',** adding, **'she came up behind him and touched the fringe of his cloak and the haemorrhage stopped that instant. Jesus said, "Somebody touched me. I felt that power had gone out of me':** *Luke 8. 43-48).*

Those were the exact words spoken by Jesus. Mark also mentions **('that power had gone out of Him':** *5.30).* This concept is of profound scientific significance, as valid knowledge, confirming that faith in the human mind has immense power of reciprocal healing: **('My daughter', He said, 'Your faith has restored you to health: go in peace').**

In November 1905, 2,000 years later, Einstein's 'Theory of Relativity', rocked the scientific establishment to its foundations, by showing that *'matter must be understood as the flow of energy',* and has continued to reverberate through all knowledge-based disciplines of science, with the revelation that all we know and perceive must be construed in terms of probabilities which have come into existence, as complex chains of interacting molecular systems, which appear to be communicating with each other along each step of the chain. Therefore, the prevailing materialistic perception of Life Processes and associated world events, can no longer be upheld. For example, the unique macro-molecular assembly of trillions of cells which make up each human being, may be understood as the emergence of the highest probability, functioning as energy gradients: (*'Today, biology is information, genome sequences are laid out* in silico *(containing silica), and Life is defined in terms of information (scientific knowledge)*...transferred by means of energy gradients*: 'The Vital Question: Why is Life the way it is?': & 'Oxygen; The molecule that made the world': Dr. Nick Lane (University College, London).*

This radical change in scientific thinking continues to inspire further advances in Therapeutic Medicine, opening the way to hitherto unimaginable developments in the treatment and cure of disease, as a new field of medical science, **'Cognitive Immunology'.** In this context, we may derive hope and comfort from past geographical experiences, which give reason to regard, what may appear to be opposite or contrary forces, may emerge as a complementary and interconnected force of immense curative power. Thus, fear of impending disaster, combined with the need to unite for the common good, tends to induce a spiritual caring quality of outlook towards other people and the world environment: e.g. The unifying therapeutic effect aroused by the *'Covid 19 Pandemic'* crisis. *(Gk.therapeutikos=healing influence: from therapeuein=treat, minister: cf. Ancient Chinese Philosophy: 'ying-yang' effect: ying= receptive : yang =active principle:* **disorder<-->order)***.

Sciences of complexity: A New Theological Resource: Cybernetics: At a time of crisis and extreme disorder throughout the Mediterranean world, the charismatic influence of Jesus Christ inspired a curative and unifying spirit of hope in the minds of all those who came to know Him, so profound that people of all religious persuasions referred to Him as, *'The Healer'.* Against all the odds, and despite the seemingly insurmountable might of Rome, this enlightened spiritual outlook enabled the formation of, *'The Christian catholic assembly of Christ'*, to emerge as a religious brotherhood, caring for each other. *('The faithful all lived together and owned everything in common; they sold their goods and possessions and shared out the proceeds among themselves according to what each one needed. They praised God and were looked up to by everyone. Day by day the Lord added to their community those destined to be saved': Acts of the Apostles: 42-47).*

The unifying power of this *Christian ethos* was eventually embraced by Constantine in the 4th Century, as the religion of the Roman Empire, exerting a curative influence throughout the Roman Empire until the 5th Century, when the division into the *'Orthodox Church '(Eastern Christian assembly),* & *'Roman Catholic Church',* coincided with the collapse of the Roman Empire, shattered by the combined effects of a long episode of migration and political turmoil throughout Eurasia, as well as lethal pandemic disease such as *'The Justinian Plague' (541-542 AD),* which killed an estimated 25 million people in the first outbreak, and over 50 million in recurring pandemic episodes during the next 2 centuries.

In the Holy land, even the Jews, Muslims and Christian assemblies managed to retain a mutual respect for each other, united in their recognition of Jesus Christ as, *('The Healer, and the greatest of the prophets': Words of Mohammed (570-632), founder of Islam);* and also by their reverence for Christ, inspired by their belief in common that these were attributes of a *divine nature.* This reasonably amicable relationship endured, albeit tinged with possessive jealousy until the ruinous onslaught of the first Crusade, ordered by Pope Urban 2, in 1095. *(Gk.ethos=cultural disposition: kharisma= favour, grace: khatholikos= general, universal belief: ekklesiastes= assembly).* Evidently, the cohesive power of *'The Christian ethos',* viewed as a human system of organization, began to weaken within a few centuries of

Christ's ministrative presence on earth, as people lost faith, depressed by the prospect of endless conflict, disease, and mounting disorder in the world around them. Theological disputes, such as the *'Nestorian schism'*, added to the growing mood of uncertainty and dismay, and faulty theosophical reasoning exacerbated the debilitating pessimistic decline. Why did Pope Urban 2 see fit to order a Crusade, inflaming hatred, and causing havoc and hideous loss of life? Wounds that continue to fester to this day. Is it possible that he could have misunderstood, *'the Will of God', as expounded clearly by Jesus?, ('But I say to you, you must love your enemies!')*

Arthur Peacocke argues cogently, that a disconnect has developed between *'Theological reasoning and interpretation'*, and human geographical experience, concerning the *'multiple interactions and connecting pathways of causality and determinative influences between entities and processes. Correspondingly, there Is also a general pressure, even among those not given to any form of traditional religiosity, to integrate the understandings of the natural world afforded by the sciences* (epistemological knowledge), *with very real, 'spiritual' experiences, which include interactions with other people and awareness of the transcendent: What Jesus Christ was to those who encountered Him and the early church* (assembly of Christ), *are about God communicating with humanity. This process of 'input of information' from God conforms with the actual content of human* (geographical) *experience, as the conveying of meaning from God to humanity. God can convey divine meanings through events and patterns of events in the created world.':* 'Information and Reality': Ch.12: pp.315-356: Arthur Peacocke). **(54).**

Thus, the myriad interactions between *'Life processes of living things'* and environmental conditions, whether propitious or deterring, can be understood as *'The Web of Life'*, through the *'Cognitive processes of the human mind'*, endowed with the mysterious abstract spiritual capacity of *'language decryption'*, communicating with the mind of God. That strong Christocentric spirit of faith, which unified the **assembly of the followers of Christ,** has been weakened and degraded by worldly self-centred interests, as well as the unchristian behaviour of religionists, like the community of nuns who presided over the *'laundry prisons'*, for the punishment of unmarried mothers.

In addition, the proliferation of *'self-styled religious sects and factions'*, including spurious ideologies masquerading under the guise of religious faith, has inflicted an even greater disjunctive effect up *'genuine true religious faith'*, and religious practice, Thus, the meaning and message of *'the word and will of God, as explained by Jesus Christ'*, has been obscured and become unclear, particularly for the younger generation and the great majority of people living in abject poverty, or struggling just to survive in a darkening world of impending decline. Paradoxically, phenomenal advances in scientific understanding, have opened the human mind, *('to see a congruence and contiguity between the nature of matter and the experiences that Theology seeks to articulate, in response to questions such as, 'What is ultimate?' One*

does not now have to choose between *'God, matter, and information', but can hold them altogether in a new kind of synthesis that obviates the false dichotomies of, Science v Humanities; matter v spirit; scientific knowledge v religiosity, that have plagued Western culture for far too long'):* Arthur Peacocke: *'Information and the Nature of Reality'.*

Yet, the confusion between *'Spirituality of Mind and Soul'*, & *'Religious Practice and Church'*, lingers on as a divisive and conflictive influence, such as the contentious strife that continues to cause such mindless destruction and loss of life in N. Ireland and the Middle East. Christ never condemned anyone accused of sinful behaviour, but was always ready to express forgiveness, except to those dissemblers who, whilst feigning to be acting in '*The name of God*', were prepared to act with indifference towards others, regardless of any harmful effects which might be inflicted upon them by their hostile selfish acts. Philosophic collective wisdom tells us that hostile behaviour generates malignant hatred in return: kindly sympathetic thinking evokes a loving response of ecological care. *'How you treat and behave towards others you do to yourself'. (Gk. sumpathes=fellow feeling: oikos=home, household: gnome=thought, judgement).* Christ expressed this even more clearly, **('Be compassionate as your father is compassionate. Do not judge and you will not be judged yourself; do not condemn and you will not be condemned yourself; grant pardon and you will be pardoned. Give, and there will be gifts for you; pressed down and running over, will be poured into your lap; because the amount you measure out is the amount you will be given back':** *Luke 6. 36-38).* Therefore, how can human beings become so mentally obtuse, as to discard the **cognitive immunological knowledge,** painfully gained from past experience, by indulging in such activities as the provocative *'Orange Order' triumphal marches in N. Ireland,* expressing continuing feelings of hatred towards fellow Christians; and the murderous conflicts in the Middle East between *'Sunni and Shiite Muslims',* which have destroyed countless lives, as well as leaving the environment in the ruinous state of a *blood-soaked wasteland.*

Christ's Challenge: The World Crossroad Dilemma: Envisage your path along Life's journey, as a road sign-posted at frequent intervals with notices, warning about conditions you are likely to encounter. Just ahead, there is a fork in the path, one way leading steeply uphill, looking rough, little used and poorly maintained; the other sloping sharply downhill, appearing to have been well-paved, smoothed and trodden by many feet, easy to follow except for a dark pool; -

--->**G. W. via 'Cybernetic Control'** : *Beware potholes: Go slow and steadily!:* **Gospel Way** -----<

----0-> **S. O. D. via 'A I' Tech. & Virtual World Gaming:** *Young people welcome: Great Prizes!*

. *'Covid 19'* **Slope of Dead-end**

Which way should we choose to go? With hindsight, we know that our fellow geographers, who have gone before us, must have faced the same dilemma. One way, looks really hard going; the downhill gradient looks more inviting, providing we can just skirt round that dark swampy patch, just past the fork in the road. We know that our forebears must have decided to take what seemed to be the easier way: but this time we have more information, haven't we? We know that from time immemorial, Buddhist theological wisdom taught that we must face both good and bad influences on our way through life, and that we can only achieve *'Nirvana' (perfection), by eschewing misleading 'desires'*. Also, ancient Chinese Philosophy teaches that achieving perfection depends upon living in harmony and ***becoming ONE*** with *'**The unfolding rhythms of the Universe': ('Thus, cells combine to form tissues, tissues to form organs, and organs to form organisms. These in turn exist within social systems and ecosystems. Throughout the living world, we find living systems nesting within other living systems':* 'The Web of Life' & 'The Tao of Physics': Fritjof Capra).

Confucius stressed that *'filial piety',* holding deep respect and veneration for our Creator and our forebears, to whom we owe the gift of **Life, is the key to peace and happiness. Christ's challenge to mankind** encompasses all of these desirable aspirations, and much more, **the way to go, ('*Tao & Faith'), in order to ensure our own future well-being, as**

well as that of generations to come. The phenomenal revelations of '*QED, Qualitative Electro-dynamics (Quantum Theory)*', have changed the way scientists perceive life and the world, even leaving Theologians bemused, unsure of exactly how to explain the unfolding '*Wonder of Creation*'.

Nothing in the Universe can now be considered wholly in material terms, as solid entities. Rather, all that we perceive and know must be understood as the emergence of the highest probability, shaped by all that has gone before, including interactions of human organization and behaviour with environmental influences. Atoms, for example, can no longer be defined as having real existence, but as phantasmal qualities of an abstruse nature, which come into existence as molecular systems, according to the meaning and significance of an inherent informational chain of instructions, as if following '*a blueprint-plan of design*', in which the '***contingent molecular systems***' seem to be able to '*converse in a logical way*', about the precise parametral changes required in each step. *(Gk.phainein=cause to appear: phantasm=unreal, illusory: metron=measure: rhythmos from rhein= to flow: parameter=feature or characteristic capable of being numerically assessed).*

Consider for example, how you see yourself, including how you are made as a macro-molecular assembly appearing as billions of interacting tiny cells, as well as the cognitive abstruse mental faculties which enable you to see yourself objectively, and to compare yourself with other human beings, including your parents and grandparents. **You are contemplating a most astounding** '*mystery of creation*'

(**'*A marvellous composition of carbon, hydrogen and oxygen atoms*',** (which do not really exist as entities), '***which have picked up a certain particular pattern in which to be arranged - enormously complicated. We cannot picture all that is really known about them chemically, because the*** *precise arrangement of all the atoms is actually known in 3 dimensions, while our picture is in only 2 dimensions'*: *'Six Easy Pieces: Atoms in Motion': 'Essentials of Physics': Richard Feynman: Pioneer of QED; 'Quantum Theory'*). **(55)**.

Scientific understanding of ourselves and human *'Life Processes',* has advanced by leaps and bounds as a result of the phenomenal revelations stemming from *'Qualitative Electro-dynamic'* discoveries at the beginning of the 20th Century, followed 50 years later by the added impetus received from decoding the language of *'The Genome': ('A synthesis of information obtained from a range of scientific disciplines, Embryology & neonatology-concerned with development, and Genetics and toxicology-valuable in identifying precise causes of congenital abnormality. Biochemistry & molecular biology are invaluable for figuring out molecular pathways following the logic of development, even down to the spatial scale of interactions between the atoms of biological molecules. Cell biology accounts for how molecular pathways are brought together to control the behaviour of individual cells. At much larger spatial scales, disciplines such as physiology, immunology and neurobiology uncover the ways in which multitudes of cells communicate and coordinate'*): *'Life Unfolding: How the human body creates itself': Professor Jamie A. Davies: Experimental Anatomy: Edinburgh)*. **(56)**.

(Gk. genos= racial type: genome=complete set of genetic informational knowledge in a cell, organelle, virus).

Thus, valuable insights concerning human development have inspired interest and research in all scientific disciplines, including mathematics, physics, computer science and the new, medically based discipline of *'Educational Psychology; Cybernetics,'* concerned with the *'Human Mind as a system of integrative control'*. In this context, Professor Davies ponders the questions arising from the mysterious transfers of vital *'Information/ Knowledge'*, by which all the interacting systems comprising *Life Processes, Cognitive Faculties, and Environmental Parameters,* appear to be connected, as if in conversation and knowing precisely what has to be done. *('How can the simple become complex? How can an error-prone system of mutually adapted parts working together construct something precise? Is human development too complex for developed humans to understand completely?'* He concludes, *'One very clear message that can be taken from our emerging understanding of development is that self-construction of a body is very different from our normal notions of construction!'* Can there be any doubt that this phenomenal enlightenment of our human mind, which commenced during the century scarred by the most destructive, hateful, and fiendish episode of inhumanity in geographical experience, is of profound significance. Viewed as an unfolding faculty of comprehension and reasoning, gathering impetus amidst the impending environmental turmoil of the 21st Century, the high probability could be that an even more alarming virulent disease may yet emerge.

The Genetic Code: Human Genome and *'Will and Word of God'*: Scientists, who only a few years ago assumed that everything we perceive could be explained in terms of material particles, such as *'bosons, leptons, quarks &fermions'*, have found themselves floundering in a *'quagmire of non-science'*, with dawning light that *'Reality is not what it seems'*: Carlo Rovelli.

Some evidently remain reluctant to discard this useless burden of nonsense, prepared to waste valuable human and physical resources in speculating on such follies as, space-travel; alien life; founding a colony on Mars; mining on the Moon; sending a probe to an asteroid; regardless of human needs here on earth. Put yourself in the place of a little child in one of the vast areas of parched scrubland in Africa, having to trudge several miles through a rough thorny track every day of her life to reach a muddy waterhole. Then, to struggle back carrying a heavy tub of water contaminated with amoeba and bacterial disease, just to keep her family alive. Imagine how the world appears to her anguished young mind. Consider how many millions throughout the world, living in conditions of extreme poverty could have been helped, if the vast resources capriciously wasted in such inept experiments, had been devoted to that purpose. *('Be patient with those who are badly off, do not keep them waiting on your generosity. For the commandment's sake go to the poor man's help, do not turn him away empty-handed in his need. Better let your silver go to brother or friend. Do not let it go to waste, rusting under a stone'): Ecclesiasticus: Generosity: 11-14).* That paradigm is the new mode of viewing the world, which is drawing the sciences together, compelling the understanding that the concept of separate groups -affiliated just by a common interest, no longer *holds water.* Rather, we are required to be united by a common *'spirituality of mind'*, as well as in our material endeavours, to counter the crises of impending ruinous disorder in our home-world, such as 'Covid 19 Pandemic'. Indeed, could that be the message which such disturbing events are meant to tell us, to know and *'see ourselves as others see us'?*

('The nature of the reality investigated by Physics and Biology is now being revealed as much more complex and mysterious than was believed. The role of information in any account of our Universe has come to take on a new importance. It seems as though the position of an entity within a structure, and the forms of its relation to other entities in that structure, call forth new principles of interaction, causing it to function as part of a complex integrated totality. New laws of nature, new ways of interaction, emerge that are not just reducible to the laws of interacting particles considered in isolation. Structure

becomes important to understanding and informational systems may be understood as having a specific function within an integrated totality that emerges only when that totality exists as a System'): 'Information and the Nature of Reality':Ch.13: **'God as the Ultimate Informational Principle':** *pp.357-381: 'The Big Questions in Science and Religion' (2008): Keith Ward, Emeritus Professor of Divinity (Oxford Un.).* **(57).**

In this way, pestilential events such as *'Corona Virus Pandemic',* assume a new and startling significance, as sub-systems of adjustment within a chain of informational knowledge directed and regulated according to, *'A Principle of Creative Control',*

God communicates with the human mind, conveying meaning and guidance, through events.

('Gk.paradeigma=example:from paradeiknunai=show side by side: from deiknunai =indicate: polymath =learned in many branches of science: poly=many + math=knowledge

Mystery of cells containing informational knowledge: The disciplines of Chemistry and Biology have been pulled together by the discovery that molecular systems, which emerge as required to underpin the existence of living things and Life processes, appear to know exactly what they are meant to do, as well as to communicate knowledge concerning previous interactions, essential to make any adjustments required to ensure the transfer of precise informational knowledge along a chain, consisting of many millions of interactions. How is it possible that every one of the billions of cells in your body, can carry all that epistemological knowledge, knowing more about you, than you can possibly learn in a lifetime? For example, *'Pyramid Cells',* discovered by Dr. Santiago Ramon at the beginning of the 19th century are now known to be, *'excitatory neurons', which carry knowledge essential to control and ensure*

precise functioning in motor reactions: *('CAGE &*
VEHICLE Cells'). This has opened new pathways of research,
which continues to open our understanding of *'the baffling*
human immunological system'. Thus, the very recent
discovery of *'T Cells, a type of lymphocyte',* which play a
central role in the body's immune system, with the ability to
kill invading virus and bacterial cells, has opened further
understanding of an astonishingly complex chain of,
'informational molecular interactions', by which the human
body seems to know how to protect itself..

(Gk. lusis= loosening; lymph from numphe= carrying water,
alkaline fluid draining into the bloodstream, which heals
inflamed tissue: apoptosis=falling away, destruction of cells).

The immune system can no longer be considered as a
relatively simple defence system, where soldier cells, such as
the beefy *'phagocytes',* are always on guard, ready to do
battle with invading infective cells. Rather, it appears to be *'a*
language learning and decoding system', capable of
intercepting the stream of molecular communication, and
devising stratagems to outwit the schemes of the viral and
bacterial invader. In this context, the really alarming
discovery about the *'Corona Virus',* is that it may have learnt
how to intervene in this *'epistemological chain of*
knowledge', by disarming the *'T cells',* which the immune
system relies upon to raise the alert by picking up fragments
of *'protein'* belonging to the bacterial or virus invader.
Studies of *'Covid 19'* have revealed that it could have found a
way to lull the immune system into a false sense of security,
by developing *'a strike protein',* which enables it to inject its
own misleading genetic information: *Chinese Study:*
DOI.1093/cid/: OUP 2020: Kings College/Barts Hospital
2020'). **(58).**

**Fig. 7: 'Geographical Quantum Fields: Networks of
Interactions':** The expression of connections linking

concepts and understanding of '*Length, Mass, Time, Temperature & Electric Charge*', as '*Quantum Effects according to the strange orderly rules of QED*'. Note the problem of reconciling perceptions of phenomena at vastly different scales.

Fig. 8: 'A Unifying Ecological View of Geographical Field Systems': Could this be the nearest we can get to an 'Ultimate Theory of Reality'?

Fig. 7: Geographical Quantum Fields: Networks of Interactions

The Biosphere: The Carbon Cycle-Photosynthesis-Respiration-(Oxidation)-Climate and Earth-habitats and biomes-Human Metabolism-Pedology and soil quality-Agriculture-Fertilizers-Food and Nourishment-Eutrification-Pathogens-Population Demograhy Distribution and Density-Urban Geography-Problems (water supply, sewage disposal, disease, education, transport)-Resources (Physical and Human)-Other living organisms

The Atmosphere: Constituents-Pollution-Emissions (Volcanic-Oceanic-Human Activity-Cattle-Industrial)-Water-(Distribution and Supply:Pollution and Health)-Stratosphere (Holosphere and Heterosphere)- Weather patterns and circulations-Shields(Ozone etc.)

Oceans: Circulations-currents-Temperature gradients-Heat store-Oceanic biology-Potential Energy Source-Weather and Climatic effects (El Nino etc.)-Ocean as a battery-Food-chains-Pollution(Oil, Acidity, Plastic waste etc.)-Ocean floor resources (minerals, methane, thermal energy)

Regional Patterns of Organization: Political Regimes of Government-FinancialSystems(Banking, investment, trading, communication etc.)-Thematic studies (Comparative studies of Human Organization (Economic models, agricultural policies etc.)-Countries-Blocs-Trade Barriers and Agreements-Development

Future Projections: Longevity-Fiscal implications-Models of Economic Development-Pensions-Actuarial Statistics-Geopolitical tensions and conflicts- Depletion of Gene Pool (wars, disease, population policies, poverty etc.)-Risk and hazard analyses

Geotectonic Solutions: Using Earth's Parameters-Deserts-Solar Power-Location of Factories of Production (water, new super-materials, prefabricated housing units, rails, conduits)-Oceans-Sources of Power(wave energy, thermal gradient energy, storage)-Chemistry of Food-Artificial habitats-Large- scale irrigation-Creation of waterways, inland lakes and seas.

Fig. 8: A unifying ecological view of geographical field systems

	Magnetoshere	
Heterosphere	Homosphere	Ozonosphere
	Stratosphere	

Atmosphere-Biosphere

Agronomics   Atmospheric Chemistry
Agriculture   **M**   Biology
Pedology
Soil Quality
Hydrography   Thermodynamics
GM Crops   Geomorphology
                                         I   Geology
                                                                        Mineralogy

               Meaning<   **>Meaning**
        > man <   **>man<** Language & Linguistics
Economics   **Purpose <**   **>Purpose**   Philosophy
Urban Design   Phenomenology
Industrial Design   **N**

Health and Disease   Psychology
                                                                       Education
                                                                       Metaphysics
                                                                       Theology

                                         D
Politics
Geopolitical Problems

Could this be the nearest we can get to the ultimate theory of reality?

'The important problems we face out there may reflect our inability to understand how the world interacts as a holistic network system'.   New Scientist: 31 May 2014: 2971.

' The end of our exploring will be to arrive where we started and know the place for the first time'
'Consciousness is where Time and the Timeless meet'   T.S. Eliot

'Principle of Universal Order: 'The Word & Will of God'

Information, Theology, and The Universe: Should this message be received as the **knell of imminent doom, or as a wake-up call?** *('The most important single issue in the conversation of theology with science is whether and how God acts in or influences the world. <u>Can the notion of information help theologians address this question?</u>):* '*Information and the Nature of Reality': Ch.14: pp. 382-404: John F. Haught: Woodstock Theological Centre: Georgetown Un.).* Dr. Haught presents a compelling argument that, '*A <u>"Universal Informational Principle "</u>, runs through all things, and that "Mind, Wisdom or Logos" inhabits and globally patterns the universe, being repeated in widely different ways, time and time again'.* In this respect, he concurs with the conviction of Mitchell Feigenbaum, one of the leading scientists of '*Los Alamos Nuclear Research Laboratory',* set up in 1942, to explore '*molecular chains of interactions',* by which energy of enormous power could be released to form an '*atomic bomb'.*

(All we perceive, know and act upon', geographical experience, must be understood as a continuous flux of dynamic change, **according to a** '*Universal Principle of Order'.* The same ontological argument was brilliantly expounded by St. Thomas Aquinas: *('Philosophia: 5 Proofs of God's existence (The Will and Word of God)':* '*Quinque Viae': Thomas Aquinas; 1225-1274).*

However, that depends on whether the human mind is listening and receptive, or dumbed down by misleading information, such as the sardonic pseudo-scientific opinion that, '*the world must have been made by "a blind watchmaker",* or by the actions and behaviour of theological and religious factions, which may be perceived to be insensitive and divisive, such as *'the Crusades', 'Orange Order Marches in Ulster',* and *'Sunni v Shiite'* destructive hateful conflicts.

Paradoxically, while scientific disciplines appear to be converging, forced into line by phenomenal revelations such as, *'The Genome expressed in the language of the genetic code',* and the significance and meaning of *'Qualitative Fluid Dynamics: Quantum Theory'-that the reality of existence can no longer be explained in material terms,* theological and theosophical philosophy appears to be at a loss to harness this new power of enlightenment,

*('**Theological ideas cannot and should not be translated into scientific propositions as attempted by "creation science" and intelligent design theory. Whenever theological speculation emulates scientific precision in specifying how divine action occurs-for example, by locating it in the hidden realm of quantum events-there is the risk of rendering mundane that which radically transforms nature: <u>faith is a matter of being grasped by, rather than of grasping, that which one takes to be "Ultimate Reality"</u>...The further theology drifts from the figurative discourse of its original inspiration, and the more closely its expression imitates the explanatory models of science, the more it loses touch with its own depth...Finally, however, our Universe, even if it is meaningfully informed, <u>is still one in which great evil can exist alongside good</u>'*): John F. Haught: 'Information and the Nature of Reality': Ch.14: Information, Theology & Universe'). **(59)**.

In this connection, geographical experience tells us that it would also be foolhardy to place absolute trust in **'Scientific Precision',** knowing that so much scientific research has been devoted to destructive and evil purposes, regardless of the consequences, as well as obscenely wasteful experiments, such as space travel, while so much needs to be done, to put things right in our home-world here on earth. In this context, Pope Francis, as a trained chemist with genuine scientific concern for the critical problems arising from rampant environmental degradation, such as 'C*ovid 19 Pandemic'*, and also as '*Pastoral shepherd of the Christian flock'*, has expressed his urgent desire for the world to unite with a healing spirit of compassionate ecological care, with particular concern for the many millions striving to survive in conditions of the direst poverty, lacking In the most basic needs such as clean, disease-free drinking water.

Thus, the Holy Father has inaugurated the *'Feast of the Sunday of The Word of God '*, first celebrated on 2nd February 2020, just before the '*Corona Virus'* disease began to escalate to the scale of a world Pandemic. There again, it cannot be regarded as just a fortuitous occurrence, that the world received an urgent ***papal message-enlightened by scientific knowledge, and inspired papal theological/theosophical wisdom counselling, 'that all Catholics should deepen their love and knowledge of God's word'***, at the same time as the startling announcement that '*Covid 19'* had become a grave Pandemic disease.

Nothing known, understood and experienced can be dismissed as a matter of chance, but must be rationalized as, ***events and occurrences shaped by systemic interactions stemming, and proceeding from the previous state of reality, perceived as geographical experience'.*** *(Gk.katholikos=universal: from katholou=in general, belief uniting a group in affiliation: ekklesiastes =member of an assembly: Hebrew; gohelet).*

In this sense, the papal message was not confined to those affiliated to religious factions attending various churches of Christian denomination, but to all mankind, ***as the 'Assembly of Christ', united in faith and belief in the 'Word of God', expounded clearly by Jesus Christ.***

Did *'Covid 19'* have the effect of uniting people, with a spiritual sense of care for the most vulnerable people, the poor and elderly? Pope Francis found it necessary, during his Mass overlooking the assembly in St. Peter's Piazza, offered for well-being of the world, to remind those clamouring for precautions to be relaxed, *'that human life was more precious than economic gain'. (BBC News: 31 May 2020).*

Evidently, there will always be those who find it expedient to qualify and obfuscate the meaning of *'God's Words',* to suit their own self-interest, such as the financial cartels promulgating large-scale gambling under the guise of promoting sporting activities, only too ready to press for resumption of competitive games and racing events, whilst *'Covid 19'* was spreading, against the medical advice that there would be a grave risk of a further peak in the death rate. *Christ castigated such guile severely: ('Brood of vipers, how can your speech be good when you are evil? For a man's words flow out of what is in his heart. A bad man draws bad things from his store of badness, by your words you will be acquitted, and by your words condemned'): Matthew 12.34-37.*

This poses a problem for all people, as geographers, naturally inclined to seek the optimum location, in order to enjoy good living conditions, and to support and ensure the well-being of our families. The challenge is to keep in mind the needs and well-being of all our fellow human beings, at the same time, and even to give them priority, according to *Christ's admonitory explanation of the Will of God.* The ancient Greek Philosophers expressed this dilemma succinctly with two injunctions: *'Know Thyself',* and *'All things in moderation':* easier to comply with in respect of friendly neighbours, but far more complicated at the scale of competitive regional and global fiercely competitive behaviour and associated activities. In this way, *'The meaning of the Word of God',* has become muddled and obscured through human frailty, with the tendency to break up established systems of social organization into like-minded factions, divided in opinion for various reasons, and as some would judge to be expedient for that purpose. We need to be reminded that the meaning of, *'Religion & religious'*, is to bind together and not to set us apart: *(Latin: religare* = bind together, as in ligature and ligament).

('If there are any wise or learned men among you, let them show it by their good lives. But if at heart you have the bitterness of jealousy, or a self-seeking ambition, never make any claims for yourself, or cover up the truth with lies—principles of this kind are not the wisdom that comes down from above: they are only earthly, animal, and devilish. Wherever you find jealousy and ambition, you find disharmony and wicked things of every kind being done; whereas the wisdom that comes down from above is essentially something pure; it also makes for peace and is kindly and considerate; it is full of compassion and shows itself by doing good; nor is there any trace of partiality or hypocrisy in it. Peacemakers, when they work for peace, sow the seeds which will bear fruit in holiness'): N.T. Letter of St James: 'Real Wisdom and its opposite': 3. 13-18).***(60).***

Young people, who turn their backs on belief in **'God's Word',** persuaded that *'Religion has done more harm than good'*, do themselves a great disservice, seemingly unaware of being mentally conditioned by misinformation, which has the effect of propagating erroneous concepts about **'God's Will and Words'.** In that way, the meaning of words as interpreted by wise minds of the past, has been twisted and subverted to suit contentious arguments, like those which caused fragmentation of the **'Christian assembly of Christ',** into churchgoing affiliations; *Presbyterian, Baptist, Lutheran, Methodist, Quaker, etc.: and even weirder cults, such as Mormons, 7th Day Adventists, & Scientology,* involving downright evil concepts, condemned by Christ in His

'*Parables*'. Thus, '*Anglo-Catholic and Roman-Catholic*', appear to be contradictory terms. Now, the forlorn empty buildings, labelled '*Church of Ireland*', set up for the hostile purpose of obliterating the worship of '*the catholic assembly of Christ's people*', evangelized by St. Patrick in the 5th century, stand neglected and dilapidated. *(Gk. katholou= universal, united in belief and faith: ekklesia from kalein= to call, summon = assembly).*

Christ's challenge to us all: Restore my Assembly!: Misinformation to be corrected and Erroneous Concepts to be eradicated:-

* *Belittlement of women and subjugation as subservient second-class citizens, sinful in nature.*

**Concept of our Creator as wrathful and vengeful, instead of loving and forgiving. The Will which gave us Life and brought all living things into being is to be loved and revered, not feared. * The World is not a dangerous place populated by hostile enemies. Divisive factions make it so. Life, living things and the state of being alive has a very special significance. We all share a world together as a family home, requiring us to bring forth a peaceful world for the good of all.*

* *Confusion has arisen between 'Spirituality of Mind and Soul', and ' Religious practice & Church affiliation'. True religious belief and practice springs from the values and store of goodness held in the heart of every man and woman, placed there by the 'Word of God', recognizable as spiritual qualities of mind, expressed especially as words and deeds of* **Love and Truth.**

It cannot be regarded as just another coincidence, that Cardinal John Henry Newman (1801-90), was canonized by Pope Francis, during the same year as the inauguration of **'Feast of the Word of God'**, in February 2020, at the same time as the '*Corona Virus*'- which had emerged in China, was perceived as a spreading virulent pandemic disease. Nothing we perceive, know and believe can now be considered as happening by chance, but as the emergence of the highest probability formed in preceding systemic interactions, just like the millions of molecular reciprocal interactions, which emerge in the form of each human being. Similarly, in 1830, John Henry Newman, (Theologian, Anglican Priest and Poet), joined with Bishop Pusey, (Theologian and Professor of Hebrew), to form and lead the '*Oxford Movement*', with the aim of restoring '*catholic spiritual vitality*', to the '*Church of England*', which- in Bishop Pusey's words, '*had lost its sacramental attraction*'.

Congregations were dwindling, and ancient village churches were falling into ruin, for lack of support and maintenance, caused b the cascading chain of destructive reactions set in train by a psychotic disordered mind, 300 years before. Understood in this way, '*The Oxford Movement*', *which ripples on,* may be construed as an immunological force, the highest probability emerging from what has gone before, instructed by '***A Universal Principle of Order***', according to the '***Word and Will of the Creator of Life***'.

The '*Covid 19 Pandemic*' may be viewed in the same way, as a corrective and cathartic reaction to clear up destructive waste from a preceding state of depredation. *(Gk.katharsis, from katharein=cleanse: from katharos= pure).*

That is '***a systems view of Life and the World as linked interactions***'. Evidently, John Henry Newman, and Bishop Pusey became convinced that they had a duty, based on their store of scholastic theological knowledge, to restore and rebuild a crumbling Church, as a material edifice which had seen better days, just as *St. Francis of Assisi* thought he was being asked to rebuild the little church of '*Portiuncula*', in a state of ruin, where in 1211 he heard the voice of Christ speaking from the Cross, *'Francis, rebuild my Church!'*. The little chapel, built during the reign of Pope Liberius (322-366), was known as *'St. Mary of the Angels'*, reputed to have been constructed of stones brought by hermits from the grave of the *Blessed Virgin.*

St. Francis began to rebuild it with his own hands, until he realized that Christ's message referred to the *'spiritual assembly united in faith and way of Life'*, rather than a material edifice where they could meet. Portiuncula Chapel is still there, incorporated now within the *Basilica of St. Francis,* in Assisi. Similarly, *St. John Henry Newman* made the same leap of *'spirituality of mind'*, when he converted to *'Catholicism'* in 1845, and proceeded to found the, *'English Congregation of the Oratory', at Maryvale,* Old Oscott, Birmingham, in 1848: recorded in his poetic prayer, *'Lead kindly light, amid the encircling gloom; Lead Thou me on!'* Are we receiving the same message, time and time again, without grasping the full meaning, so misinterpreting the true significance of *'Christ's Word's'?*

Our Confused and Deranged World: The first few years of my life are remembered as a time of prosperity and happy family life. My father had worked hard, up to 16 hours a day from the age of 14, not only continuing to assist in running the *Greengrocery shop,* established by his father, but also working as manager of a watch and clock Company, set up in London by a *Russian Jewish Emigre,* who had managed to escape the pogrom in Russia. Thus, with my two brothers, we were able to enjoy a family holiday in Ireland every year, long before it became affordable and customary for *working class families* to take an annual holiday. My father believed that it was of paramount importance, as a duty of care, to preserve and strengthen the bonds of *loving family relationships,* and that we should know and respect our grandparents. Then, in 1931, our family life was shattered, when *'The Great Depression'* burst upon the world, and my father, with a week's notice, was thrown out of work, bringing our family to the brink of starvation. At the age of 6, my perception of the world around me changed. It became a place fraught with danger, instead of the happy propitious environment I had come to know. That was the beginning of an episode of dark travail on my way through *'Life's journey'*, which became a menacing experience of *'encircling gloom'*.

In November 1931, my older brother, Andrew aged 7, died as a result after a fall and head injury. With hindsight, he was a most unusual child; very peaceful and loving, especially to our parents; with an extraordinary sense of religious devotion. From the age of 6, he began to serve at the 8 a.m. Mass, in the Dominican Church next to our School. I had to go with him, walking over half a mile from our home, starting about 7-30 in the morning. He had taken me to school, holding my hand since I reached the age of 3½, because in those days a child aged 5 could take a younger sibling with him. Even when it was stormy and pouring with rain, his mind was set on serving at Mass. I recall that as he was dying, Fr. Dominic came to our home, and I heard him say to my sobbing mother, as he was cradling Andrew, *'Andrew is a very gifted soul'*.

As a result, my poor mother lost her reason: for months on end, I would find her standing at the window, crying and tapping at the glass, unable to prepare meals or to look after my younger brother aged 4, who then became seriously ill, with convulsive fits, arising from severe constipation, caused by an inadequate diet consisting mainly-as I recall, of stale bread and over' ripe banana mashed up in milk. How long this downhill chaotic experience lasted, extending over several months, has become clouded in amnetic haze. *(Gk..amnesia=forgetfulness: Khaos= vast chasm, yawning gulf, formless; strepto=twisted: kokkos =seed, bacteria).*

The next sharp memory, with any clear sense of childhood, is of being in a strange condition of isolation in a grim hospital ward, where I found later that I had been quarantined with the highly contagious disease of scarlet fever, caused by streptococcal infection, at the age of 7. Curiously therefore, my awareness of being in a state of childhood remains sharper at the ages of 3½ and 7, than in the hazy episode between. I could not have expressed this experience of *'being in a state of probabilistic existence'*, at that time: but, I now regard it as the first inkling that *'reality may not be as we perceive'*, rather an opening of mind to understand that all we experience, *"reality of geographical interacting systems'*, may be viewed as in a state of *'fluid dynamic change'*.

Nothing can be considered as *'solid and fixed in a state of material existence'*. Even the chains of *'molecular messengers';* informing the cellular systems; which enable you to read; decode the language of information you are receiving; make sense or otherwise of the knowledge gained; enabling the higher *'Cognitive Processes-logic and spirituality of mind'*, to steer and guide the way of transition unfolding as *'highest probability to emerge'*, must be viewed as in a constant flux of change. Therefore, what appears as material and corporeal must be understood as the transition of a formless state of probability, emerging in a structured state of form, according to *'a universal principle of ordered organization'*. For example, a unique human being like you and me, an astonishing macro-molecular assembly endowed with *'life systems'*, as well as *'systems of cognitive capacity'*, to know and understand ourselves and the ambient world systems supporting us.

Plato displayed extraordinary precocious scientific insight in his dialogues concerning form, preoccupied with the meaning and significance of shapes, expressed in his *'Theory of Forms'*. This postulated an *'Ideal World Order of Forms (Organization)'*, transcendent to the world we perceive, *'which is defective and full of error'*. He would no doubt have been intrigued, to find that 2,500 years later, the questions he had raised, were assuming greater significance, drawing all scientific disciplines together, by the revelations of *'QED (Qualitative Electrodynamics: Quantum Theory)'*, showing that atoms must be considered as abstract *'probabilities'*, which emerge as molecules and molecular systems, apparently imbued with *'information' instructing what they have to do'*.

Liquidity of Water: H^2O: Why should this simple molecular system be endowed with this strange slippery quality? Plato pondered this question, postulating that *'water must be the spring of Life'*. Water certainly has mysterious properties, just right for Life to exist. We now know that liquidity can be explained as, *'weak force of the Hydrogen bond'*, as if 2 Hydrogen atoms are engaged in a dance with a heavier Oxygen atom, like 2 nimble girls, fleet of foot, dancing around a heavier and clumsier male. But, even more mysterious properties, essential and just right to support Life processes, remain to be explored. For instance, recent research has revealed that water appears to have over, *'70 anomalous chemical properties'*, and 200 different fluid structural patterns have been observed.

-- In lakes and seas, water never freezes to the bottom level, even in the Antarctic, where the lowest temperatures have been recorded on Earth, leaving room for living organisms to survive; --- In the microtubules of the brain and kidneys, thousands of times finer than a human hair, water seems to know that molecules have to form up in a single line in order to flow through, in order to do their work. Otherwise, the kidneys could not purify bodily fluids, and we would not be here. Long before his time, it appears that Plato was able to perceive and envisage the world as connected and interacting systems, at different levels of scale, down to the minutest, without the empirical scientific knowledge we now have. Thus, it is evident that he was able to contemplate the dilemma presented by the state of our world, *'deranged and full of error'*, instead of the *'ideal World'* it is meant to be, regulated according to a *'a basic principle of ordered organization'*, that we all desire. It seems that we are receiving the same message from discerning spirituality enlightened minds, again and again, without paying much heed to it: e.g. ***Thales 6th BC; Plato 4th BC; Judaic Scriptures; Gospels of Jesus Christ; Francis of Assisi 12th AD; Thomas Aquinas 13th AD; St. John Henry Newman 19th AD; and Pope Francis 21st AD.*** We are still perplexed by the worsening dichotomous state of the world that engaged the philosophers of ancient Greece, seemingly unable to unite, to ameliorate the deteriorative effects arising from the problems we are facing, such as *'Climate Change'*, and *'Covid 19 Pandemic'*.

(Gk dilemma= assumption: dikhotomia=division, split world: krisis fom krinein= decide, turning point: problema from ballein = throw, put forth, arisng, connected with).

(*'Only faith can guarantee the blessings we hope for, or prove the existence of the realities that at present remain unseen. It is for faith that our ancestors were commended. It is by faith that we understand that the world was created by one Word from God, so that no apparent cause can account for the things we see'):* St. Paul: Letter to Hebrews;11. 1-3. **(61).** Surely, the same message as that conveyed by the revelations of *'Quantum Electrodynamics!*

The Dilemma of a dividing and fractious World: How can this state of torpid ineptitude be explained? In spite of the knowledge gained from phenomenal scientific discoveries during the last 50 years, concerning the web of interactions by which, *'Life Processes'* appear to be closely interlinked with *'Environmental parameters'*, *(Geophysical Systems)*, it does not become easier at world level, to reach agreement in making wise decisions, designed to counter the problems confronting us: *'**United by doubts but divided by convictions**

and self-interests'. A new insidious *'Pandemic of mental psychosis'* is creeping invisibly across the world, a severe mental disorder involving a loss of contact with reality, with the perfidious effect of dimming the higher cognitive faculties of the Human Mind. Far more lethal than *'Covid 19'*, the symptoms are plain to see: increasing learning disorders, *(dyslexia, dyscalculia, dysgraphia; Attention deficit/ hyperactivity disorder-ADHD) etc.);* addiction to *'virtual-world gaming'*-associated with gambling; fear of a dangerous world and hostile people, leading to self-harming, and loss of *' will to live'*, with an increase in the rate of *'juvenile suicide'*.

(Gk. Psukhe =breath, life, soul, mind; Lethe= resulting in death, causing or culminating in spiritual oblivion).

The knock-on effects are even more corrosive, by splitting society relationships across the world into antagonistic factions: e.g., between the younger generation and their forebears; the different ethnic groups; the educated professional class and lower paid workers; the East against the West; the Police and the law-breaking criminal class; the ultra-rich and the down-trodden poor; varying religious and ideological beliefs. In this way, a spirit of envy, jealousy and hatred is being cultivated as a venomous divisive force, replacing the **God-given bonding empowerment of 'Love and Compassion'.**

(*'By His divine power, He has given us all the things that we need for life and true devotion, bringing us to know God Himself, who has called us by his own glory and goodness. In making these gifts he has given us the guarantee of something very great and wonderful to come. Through them you will be able to share the divine nature and to <u>escape corruption in a world that is sunk in</u> vice. But to attain this you will have to do the utmost yourself, adding goodness to the faith that you have; understanding to your goodness; self-control to your goodness; patience to your self-control;*

true devotion to your patience; kindness to your fellow men to your devotion; <u>and to this kindness, Love</u>. If you have a generous supply of these, they will not leave you ineffectual or unproductive: <u>they will bring you a true knowledge of our Lord Jesus Christ'</u>: 2 Peter: 1-7. **(62)**.

(Gk. antagnizesthai= struggle against, hateful enemy; ethnos= nation, different group).

Warning to the younger generation: Do not be fooled by friendly, plausible *'Uncle S. and bewitching Auntie L.'*, into joining the new *'Idolatrous Cult of Tech.'*, paying homage to the *'Idol of 'Robotic AI'*. The target is your *Human Mind,* the most valuable immunological defence system we have, to steer us safely through the *swamps of error and vice,* on Life's journey.

(Gk. kubernetes= steersman from kubernan= steer).

Cybernetics: Science of 'Control of Mind': *('Never say, "God sent the temptation", when you have been tempted: God cannot be tempted to do anything wrong, and he does not tempt anybody. Everyone who is tempted is attracted and seduced by his own wrong desire. Then the desire conceives and gives birth to sin, and when sin is fully grown, it too has a child, and the child is death.... Where do these wars and battles first start? Isn't it precisely in the desires fighting inside your own selves? You want something, and you haven't got it; so you are prepared to kill. Anyone who can bring back a sinner from the wrong way he has taken will be saving a soul from death and <u>covering up a great number of sins</u>'):* The Letter of St. James. This guiding wisdom is as pertinent today as it was 2,000 years ago. It is not God *'who leads people into temptation'*: the younger generation is being led astray in so many ways, by misleading information and erroneous concepts, such as, *'Religious ideas are old-fashioned, we live in a secular world',* which must be *'anathema to our Lord God, who gave us Life and brought us into being'.*

The wave of bitterness and conflict which we perceive to be mounting, and sweeping throughout society relationships, aided and abetted by the loud clamour of misleading information, does not bode well for the future. Our home-world may be poised yet again, at the threshold of a steep decline into, *'a vale of darkness and trauma'.* It would be well to remember that what is about to happen, has been shaped by all that has gone before, and that anything *'destructive and deathly',* **which may emerge cannot be attributed to** *God, our Creator,* **but to what has been stored in the Human Mind.** That is the same dilemma, which preoccupied **St. John Henry Newman and Bishop Pusey,** in 1835, when they founded **the Oxford Movement, with the intention of restoring faith in God, as it was in** '*The assembly of believers formed by Christ'.*

(Gk.anathema, from anatithenai=to set up: Maran atha from Aramaic **'maran ata'='Come Lord to our aid',** in order to combat and overcome these destructive influences.

The Oratorian Ethos:
Cardinal John Henry Newman

The Oratorian Ethos: When my family stumbled into a '***vale of darkness***', in 1931, engulfed in a cloud of gloom and grief-stricken misery, I could not have foreseen any possibility of escape. Then, in 1935, the way out opened suddenly, when I was called -at the age of 10, into the headmaster's office in St. Dominic's Primary School, Hampstead. I had no idea what to expect. I had been offered a place in London Oratory School. My view of Life and the dark world which had surrounded me for 5 years, half of my childhood existence, was changed in a trice. Hope really does '*spring eternal in the human breast*', in response to the saving grace of a caring thought, and a helping hand. Stimulated in this way, the Human Mind can become further cognitively reactive, as if catalytically energized.

(Gk. Katalusis from kataluein: luein= set free, get going, cause a rapid instantaneous reaction).

The 'Prayer of St. Francis', asked for that curative power of grace: ('*Make me a channel of Your Peace, Where there is despair let me bring hope'*). Privileged to take up my place as a pupil in London Oratory School, in 1936, that was how my mind opened to perceive everything, and to know myself in a completely different way, reassured with restorative hope. That benign influence proved to be a pillar of strength during the 2nd World War, on occasions darkened by stress and spiritual gloom. At the end of war, in 1946, the Secretary of State for Education, requested that all qualified teachers serving in the Forces should be released, to meet the desperate shortage of teachers. The 'All-age Schools', many dating from Victorian times, were in a state of wretched decrepitude. Children were dropping out of school before the age of 14, or 12 in rural areas. Many teachers were still working, long past the age of retirement. Thus, I found myself unexpectedly released from the RAF, with one day's notice, in search of a teaching post. Drawn by a nostalgic memory of my schooldays as a time of, *'homely well-being'*, and as I was near Stewart's Grove in Chelsea, on a sudden impulse I decided to call in to my old school, where Joe Kelly, my old Geography teacher, was acting as Deputy Head. On hearing that I was looking for a teaching post, he asked to see my Teaching Diploma.

To my utter astonishment, noting that I was completing an Honours Degree in Geography, with ancillary studies in French language and Literature, as well as the Geography and Geology of France, Joe asked me to join the teaching staff. So, I was able to commence my teaching career in my old school, as Form Master in the very room, now in a tattered state, where I had sat in a dazed state of elation, 10 years before.

The Oratorian Tradition: The Oratorian Movement and way of life, is essentially based on prayer, rather than a system of rule or regulation. *(L. Oratory=place of prayer: from orate=pray, plead)*. It has grown from the seed of ***'Faith in the Word and Will of God'***, which took root amongst the sick and poor in 16th Century France. Following the *'Black Death Pandemic'*, which killed about one third of the peasant population during the early 14th Century, the series of conflicts, which came to be known as, *('The Hundred Years War' (1337-1453)*, had reduced France to a ruinous state, shattered with dissonance and disorder. Five generations of royal dynasties, involving warfare between the *'House of Plantagenet in England'*, and *'The French House of Valois'*, had fought continuously to rule over France, the largest Kingdom in W. Europe. The cumulative effects of depredation and loss of life upon the working peasant population may be imagined. By the early 16th Century, the hideous *'gap of inequality'*, between the aristocratic ruling class, living in

147

luxury, and the poverty-stricken working population, riddled with disease and festering from wounds of war, was becoming unsustainable. Life must have appeared very bleak indeed for the common underclass in France, at the beginning of the 16th Century, with no hope of escape from the misery of unaided poverty, '*a vale of darkness*', closing in around them. Yet, that was when a glimmer of light appeared, in the helping hand of a saintly priest, who dedicated his life with missionary zeal to labour among the sick and poor.

St. Philip Neri (1515-1595): Canonized in 1622, St. Philip Neri became known as, *'The Apostle of Rome'. (Gk. apostollus=messenger: from apostellein= to send forth, spread the Word of God).* 'The Congregation of the Oratory', founded through his initiative, gathered divinely inspired priests, to pray and work together, exerting an immunological influence of restorative healing. Whatever happens, all that we know and perceive through geographical experience, emerges as the highest probability shaped by previous interactive influences. If the dominant influence stems from the storage of goodness in the Human Mind and heart, emergent systemic influences will trend to unite and heal discord and suffering. Conversely, the effects of a greedy aristocratic ruling class, living in extreme luxury, with blatant indifference to the sufferings of the downtrodden poor, will inevitably evoke hateful resentment, leading to riotous rebellion, which may ripple on for centuries to come. Similar conditions of obscene inequality existed in 16th Century England.

Hampton Court Palace bears witness to the life-style of the Tudor Court, including Cardinal Wolsey, regarded as a *'Prince of the Church'*, who built it as his home in 1515, suitable for his sumptuous style of living. No wonder, that the *'Catholic Assembly of Christ'* came under attack, led by a psychotic despot, who found it an obstacle to his insatiable desires. The way was paved for the brutish onslaught of the next tyrannical Dictator, in the guise of *'Reformer'*, but who gathered all those crippled, mentally deformed, and unwanted as a burden for any reason- including women, consigning them to the W. Indes as indented *'white slave labour'*. So, as the rot sets in, it opens the way to further decay, turmoil, and mounting disorder. Do you recognize these signs?

('Among pagans it is the kings who lord it over them and those who have authority over them are given the title benefactor. This must not happen with you. No! The greatest among you must behave as if he were the youngest, the leader as if he were the one who serves'): Luke 22. 25-27.

It would be mistaken, to regard the ***'Word of God',*** as just a passive message. Rather, it must be understood as a vibrant spiritual empowerment, a reactive corrective influence, always waiting in the form of *'Wisdom'* for any one worthy to receive it. To those who say, *'Why does God let unpleasant and dangerous things like 'Covid 19' disease happen,* the answer was given by Christ to Pontius Pilate, Roman Governor of Judaea, who asked, *'Are you King of the Jews?.* *('Mine is not a kingdom of this world!': John 28-40).* Also, when the Pharisees asked when the kingdom of God was to come, Jesus answered: *('There will be no one to say, "Look here, look there!" For you must know, the kingdom of God is among you': Luke 17. 17-21.*

Our Creator was telling them the simple truth, as confirmed by the revelations of 'QED (Quantum Theory), that it is illusory to think of Life and geographical experience in a world remarkably ordered to support *Life Processes,* in material terms. The Sciences have also been shaken to their foundations by the discovery that **nothing in the Universe can now be considered as materially solid,** having to resort to *'Theological Philosophical'* reasoning of old, to explain mysterious interactions, such as the *'ordered macro-*

molecular assemblies', which emerge in the form of each unique human being, and every living organism. Are we not being made to see that understanding of *'God's Word and Will'*, comes through events and patterns of events in our created world, perceived as regulated by, *'A universal principle of order'?* Thus, amidst the hopeless conditions of mayhem, turmoil and suffering in 16th Century France, the *'Oratorian Ethos'*, inspired by St. Philip Neri, took root and flourished.

St. Vincent de Paul (1581-1660): The next saintly priest to rise to Christ's challenge was renowned for his compassion, humility and generosity, and especially his loving care tor the poorest and homeless down-and-outs. He dedicated his life to serving the poor and needy, working tirelessly day and night, and was revered for his *'spirituality of intensive love'*, radiated by his unremitting efforts to alleviate their sufferings. He founded *'The Congregstion of the Mission and Daughters of Charity of St. Vincent de Paul'*, and was canonized in 1737. That is how *'God's Will'* may be seen to have been actively working, from the beginning of Creation, as a curative power of well-being for the good of the world. A spirit of loving care, begets respect, energizing the higher cognitive faculty of aesthetic appreciation in the human mind, which in turn gives rise to a hopeful optimistic outlook, instead of downcast pessimism. Hope paves the way for the soaring empowerment of *Faith, which engenders the charisma of Love and Truth in dialogue with God's Holy Mind: 'Deus Vult!':* (God Wills it so.).

('From a Theological perspective, there must also be an ontological connection between the world, by virtue of the incarnational structure of the doctrine of creation: God is present in the midst of the world of nature as the informational principle (Logos) and as the energizing principle (Spirit)...What is ultimate reality from a physical point of view is penultimate from a theological perspective'):

'Information & Nature of Reality': Ch.15. pp.405-443:'God, Matter & Information':Gregerson,N.H.: Copenhagen Un). **(63).**

Those who, without thinking, spread the myth,*'we live in a Godless secular world',* do themselves and their offspring a great disservice, as Blaise Pascal (Mathematician of *'Probability Theory'),* replied to the student who asked about the probability of there being a God. The choice is ours. Closed minds bring forth a mood of fear and pessimism, leading to hateful antagonistic relationships, and ultimate destruction. A caring responsive spirituality of mind, unites all in a bond of mutual brotherhood, with the capacity to think and act wisely, ***'in order to bring forth a pleasant happy home-world for the good of all living things, which we share together',*** according to ***'Will and Word '*** of our Creator. ***('The presence of the resurrected Christ, conveys the powers of love, forgiveness, healing, and His passion for children, the weak, the rejected, the sick, and those in misery. Further, the power to confront the so-called 'powers and principalities' begins to take shape in His presence: for example, in conflict with political and religious institutions in the search for truth and demands for real justice'):*** *'What may be regarded as 'ultimate' in the interrelation between God, matter, and knowledge: Ch.16: pp. 444-462: 'Information & the Nature of Reality': Welker, M.: Heidelberg Un).* **(64).**

Blessed Frederick Oznam (1813-1853): Frederick Oznam was born during the militaristic regime of the Napoleonic Era, following the calamitous horrific events associated with the *'French Revolution'* which broke out in 1789, when all the women working in the Paris markets rose up as an angry aggressive mob, and marched to the Palace of Versailles where Louis XVl and Marie Antoinette with their aristocratic courtiers were living in voluptuous luxury. By contrast, Paris had become a cesspool of lawless iniquity, riddled with disease and crime, with the population reduced to a state of ravenous desperation. The Louvre had been abandoned as a royal palace in 1682, when Louis XV moved with the Court to the new Palace at Versailles. When Marie Antoinette was informed by a courtier that the people *'had no bread'*, and were dying from starvation, it is recorded as her flippant response,*"Donnez-leur de la brioche"*: *Oh, give them some cake'*. Whether true or not, the people could no longer contain the corrosive mood of hateful resentment, inflamed by the flagrant disparity between the luxurious lifestyle of the aristocratic rulers, and their state of abject destitution. ..*('Gk. aristokratia, from aristos= best: aristokratikos= rule by the best and most worthy citizens)*.

Furthermore, the prelatial dignitaries of the Catholic Church, *'Christ's Assembly'*, had become embroiled with the political agenda and militaristic wranglings, between the warring European Royal dynasties, which launched the *'Protestant Reformation'*, leading to the fragmentation of the Catholic Church.

When Christ's admonitory words fall unheeded on deaf ears, the meaning and significance of **'Christ's resurrected presence' also becomes obscured, and blotted out'**, in *'Christian Liturgical,* worship. **('Put your sword back, for all who draw the sword will die by the sword'):** *Mat. 26. 52-53:* & **('You must love your neighbour as yourself")**: *Mark 12.31-32. (Gk.leitourgos=public worship: eukharistos from kharisesthai=giving thanks)*.

Thus, the people of France, disillusioned by the unchristian behaviour of ruling prelates, in collaboration with a corrupt ruling class, rose en masse in the French Revolution, to declare France as a *'secular State'*. **('La France et une Republique indivisible, laique, democratique et sociale'):** *Article 1 of the French Constitution.* Thenceforward, religious involvement in political affairs would be actively discouraged, thereby excluding consideration of the teachings of Christ as a salutary influence for the good and well-being of mankind. In that way, the pent-up anger of the enraged mob boiled into bestial hatred, culminating in the hideous public spectacles of mass execution by guillotine, of the royal family and all those associated in any way with the aristocratic ruling class, as well as property owners and anyone seen to be comfortably off. For example, both Antoine Lavoisier and his wife Marie-Anne, famous for their Chemical discoveries in Paris University, were executed by guillotine in 1794, without mercy just for being rich, whilst *'Les Tricoteuses'* (women knitting), jostled to get a good view, screaming with delight, as heads rolled. We are reminded of similar macabre patterns of geographical experience, following enforced regimes of godless secular rule; the Mongol depredations; Tudor psychosomatic atrocities; genocidal tribal warfare in Africa; the *'Beast of Belsen'; 'Pol Pot' cleansing'* of the professional and educated class; the Chinese Mao Tse Tung *'Cultural Revolution'*, etc.

The dystopian effects of these diabolical episodes of depravity continue to influence whatever may emerge, as the *'highest probability'*, of systemic development.

At the beginning of the 19th Century, when it must have seemed that human behaviour had fallen to an abysmal nadir of depravity, Frederick Oznam appeared, as a light amid the dark gloom of despair cast by the excessive cruelty witnessed in the previous decade. From a family of Jewish origin, he worked tirelessly to mitigate the sufferings of the poorest and weakest, including those from prosperous families burdened with grief for loved ones summarily executed during the Revolution. He became revered as a loving holy priest, founding *'The Society of St. Vincent de Paul'* in 1833, an International Voluntary Organization, which has grown as a system of organization to assist any found to be in need, in every Christian assembly throughout the world. He was

'Beatified' by Pope John Paul in 1997, and his beatific loving influence continues to gain impetus, showing that the spirituality of love *'within the store of goodness'* in the human heart, not only exalts the senses, but also works as an empowerment of gracious behaviour, mindful of the needs of others. Whereas loving deeds evoke a mutually grateful and loving response in the hearts and minds of those we care for, duplicitous double-dealing sows the seeds of contempt and hatred that have generated the most heinous acts *'of inhumanity to man'.* Thus, the weakening of faith in Christian religious worship, as well as the secularisation of France, in the French Revolution, can be attributed to the readiness of Catholic dignitaries to conspire with ruling governments in their machinations against each other.

For example, Voltaire (1694-1798), noted for his sardonic satirical outbursts, campaigned bitterly for the separation of Church and State, arguing that the Church should not be involved as an enforcer of government: *('It is forbidden to kill. All murders are punished, unless they kill in large numbers, and to the sound of trumpets':* *'The art of government is to make two-thirds of a nation pay all it possibly can for the benefit of the other third').* He was not an atheist, as many would aver, but took an extremely jaundiced view of prelates who were apparently willing to serve aristocratic ruling cadres in their various intrigues, while showing little regard for the plight and sufferings of the poorest and weakest people. He became obsessed by what he regarded as their *'sanctimonious hypocrisy, talking of holiness and practising injustice'.*

We may be reminded that Christ hardly ever spoke in words of anger and condemnation, as recorded in the Gospels. On the contrary, when confronting human frailty and sinful behaviour, He always showed readiness to forgive with words of gently chiding sympathy, except when facing the *'hypocrisy and vanity of the scribes and Pharisees'. ('They*

154

occupy the chair of Moses, You must therefore do what they tell you and listen to what they say: but do not be guided by what they do; <u>since they do not practise what they preach.</u> They tie up heavy burdens, and lay them on men's shoulders, but will they lift a finger to move them? Not they! Everything they do is to attract attention. Alas for you scribes and Pharisees, you hypocrites!... You who are like whitewashed tombs that look handsome on the outside, but inside are full of dead men's bones and every kind of corruption. In the same way you appear to people from the outside like good honest men, <u>but inside you are full of hypocrisy and lawlessness')</u>: Matt. 23. 1-32.

(Gk.hypocrite=dissembler,pretender;fromkrinein=determine, judge: sumpatheia from sumpathes=having a fellow feeling).

Jesus also warned that we cannot serve, *'both God and Mammon',* with reference to the love of money as an evil influence, in the greedy pursuit of wealth and power. *('No servant can be the slave of two masters: he will either hate the first and love the second, or treat the first with respect and the second with scorn. <u>You cannot be the slave of both God and money')</u>: Luke 16. 13. (Gk. mamones from Hebrew, Aramaic= idolatrous devotion to the acquisition of riches).*

John Henry Newman (1801-1890): The same dilemma was addressed by two Theological scholars at Oxford University in 1835, Bishop Pusey (Professor of Hebrew), and John Henry Newman (Anglican Priest and Theologian), who founded the *'Oxford Movement',* for the express purpose of *'restoring vitality to liturgical worship',* in the Church of England. They had become dismayed by the decline in church attendance, and the general fall-away in religious faith. Even the ancient village churches were falling into ruin, for lack of interest and care. The wedge of division, driven between the *'hallowed British Christian Assembly'* and the Papal authority of Rome, caused by the vicious Tudor pogrom, was continuing to obscure *'the significance of the Lord's sacramental presence in the Eucharist'*. The age-old message of Holy Mass, in the words of Jesus Christ at the *'Last Supper'*, (*'Do this in memory of Me'*), had lost its meaning as a *'Feast of rejoicing in Christ's resurrected presence'*. They were appalled by the effects of the spreading nihilistic philosophy of secularism, excluding concepts based on belief in God and a future existence. In one of his sermons (1843), John Henry Newman, expressed their dismay about the loss of faith in the *'Church of England'*, by spelling out what it means to be a member of *'Christ's assembly'*: *'A Christian looks for Christ: not for gain!'*. Both of them shared a sense of scholastic acumen, together with a wealth of epistemological knowledge, which enabled them to view this progressive spiritual decline in geographical dimensions of space and time.

With the advent of the Industrial Revolution, young agricultural workers were migrating in droves to the towns, disgruntled by their low pay and servile conditions, and drawn by what appeared to be a better way of life, working as servants in prosperous families, and much higher wages on offer to industrial workers.

As a result, the cohesive bonding of a rural community, centred on the Christian spirituality of the village church, was being replaced by a materialistic outlook driven by monetary gain. The *'inequality gap between the haves and the have nots'*, was widening apace, as the new industrial entrepreneurs became vastly rich on the backs of an unlimited supply of cheap labour, willing to work in extremely dangerous and insalubrious conditions, such as coalmining and iron and steel foundries. As devout *'Anglo-Catholics'*, Bishop Pusey and John Newman looked back with abhorrence at the desecration and fiendish destruction wrought by the psychopathic tyrant who founded *'The Church of England'*, to serve his own insane desires, when the altar tables upon which '*The Holy Mass in memory of Christ*' had been celebrated for centuries, were torn down and set in the floor. Bishop E. B. Pusey, in the context of his understanding of '*Christ's words-spoken in Aramaic*', refers to such acts of profanity by analogy: **('The vessels of the Temple, desecrated by being employed in idol worship').**

They longed for these heathen profanities to fade into insignificance, hoping for the time when, **'Christ, the higher gift than grace "God's very presence and His very Self, and Essence all-divine",** would be reinstated in His rightful place, among His '*Catholic Assembly*', reconsecrated at the head of the table, and all would again be reunited in the bond of '*Christ's Love*'.

In 1845, John Henry Newman, made a great leap of **'faith'**, recognizing the pernicious influence of materialism as a symptom of '*Mammonism*', the lurking idolatrous malaise of old, emerging again in a more virulent form, as a '*Pandemic of psychotic disorder*', inflamed by an inordinate lust for power and wealth, in direct contravention of the **'Teachings of Christ'. ('Before Christ came was the time of shadows but when He came, He brought truth as well as grace; and as He who is truth has come to us, so does He in return require that we should be true and sincere in our dealings with Him'):** *Excerpt from Advent Sermon: 'Unreal Words': Quoted and explained in Ch.3:'Watching for Christ': pp.27-39:'Newman: The Heart of Holiness': Prof. Mon. Roderick Strange: St. Mary's University).* **(65).**

His soul-searching contemplation of the divinity of Christ, **'The Essence all-divine',** diminished and obscured in the so-called '*Protestant Reformation*', made him realize that the very '*presence of Christ*', was no longer acknowledged in the liturgical practice of The Church of England.

('The course of error then at the present day seems to be this-- first to forget that Our Lord is the Son of God in his

divine nature –to speak of Him vaguely as God, which most surely He is but vaguely, not as being what He is also-- God from God –Light from Light – very God from very God'): 'Newman: The Heart of Holiness': Ch.4: 'Life in Christ': pp.41-56: Mon. Roderick Strange). **(66).**

When He finally made up his mind to convert back to 'Catholicism', in 1845, announcing his acknowledgement of the true presence of Christ in the 'liturgical worship of Holy Mass', he later described what he felt as like 'coming into port after journeying through a stormy sea, in perfect peace and contentment, without a single doubt'. However, this did not please many Anglicans who, although of similar mind, could not bring themselves to confront the venality of the cosy relationship established between State and Church. For example, Bishops of the Church of England were elected to the House of Lords, as required by their 'raison d'etre', to lend credence to the legality of State enactions. Jesus Christ gave warning of this when referring to His second coming.

('Take care not to be deceived, for many will come using my name-- Refuse to Join them-- And when you hear of wars and revolutions, do not be frightened, for this is something that must happen, but the end is not so soon. Nation will fight against nation, and kingdom against kingdom. There will be great earthquakes and plagues and famines here and there; there will be fearful sights and great signs from heaven. But before all this happens, men will seize you and persecute you --- that will be your opportunity to bear witness – your endurance will win you your lives'): 'The Warning Signs': Matt. 24. 4-14).

As a result, John Newman was subjected to a torrent of vilification and abuse, and many rumours were circulated about his state of mind, alleging that he was unstable, disloyal, unsure of himself, and even planning to revert to the Church of England, but he kept staunchly to his decision: *('I* *do profess that Protestantism is the dreariest of possible religions; that the thought of Anglican service makes me shiver, and the thought of the Thirty-nine Articles makes me shudder. Return to the Church of England! No; 'the net is broken, and we are delivered '. I should be a consummate fool (to use a mild term), if in my old age I left 'the land flowing with milk and honey' for the city of confusion and the house of bondage'):* 'Newman: The Heart of Holiness: Ch.7: 'In Darkness': pp.81-97: Mon. Roderick Strange.* **(67)**.

The false rumour spread that his conversion ended the bond of intellectual brotherhood between him and his old friend, Bishop Pusey, but this was not true, as I discovered for myself when I visited the Anglican Convent of the *'Sacred Heart',* in Fernbank Road, Ascot. My wife had been a boarding pupil there, when it was an Anglican School for girls, from 1942 to 1944, after she was left an orphan during the first years of WW 2. We had gone to visit the only surviving Nun, Mother Superior. During our discussion, she mentioned that Pusey had been devastated when John Newman had suddenly announced his conversion to the *'Catholic faith'* but did not leave us with the impression that they had parted as enemies, as was being rumoured in the newspapers, in 1865. Whilst she was not well pleased to learn that Mary had herself decide to convert back to Catholicism- her Mother's faith, when she later attended St. Bernard's Convent School in Westcliff, where she formed a lasting friendship with my two sisters. She was sure that Pusey and Newman had remained good friends, with deep and loving respect for each other, united in intellectual brotherhood for the rest of their lives.

('The deep love between us, which now dates back over 40 years, has never been in the least overshadowed. His leaving us was one of the deep sorrows of my life, but it involved separation of place, not diminution of affection'): Letter to Guardian: E. Pusey: 'John Henry Newman: A Mind*

Alive':Ch.7: 'Seeking Church Unity':pp.93-108: Professor Mon. Roderick Strange: St. Marys University). **(68).**

Edward Pusey had been born as, Philip Bouverie, the younger son of Jacob des Bouverie, of a notable aristocratic Jewish family, 1st Viscount Folkestone, and had inherited the Family Estate. He was educated at Eton, became a Commoner of Christchurch (Oxford), and was elected to a fellowship of Oriel College, where he met John Newman and John Keble. His close associates at Christchurch College, included Lord Carnarvon (4thEarl;1831-1890), who later was twice Secretary of State for the Colonies, and Lord Lieutenant of Ireland. He had a son, Philip Edward, (1830-1880), a noted Hebraic authority, following in his father's footsteps, who edited an edition of St. Cyril of Alexandria's *'Commentary on the Minor Prophets',* and three daughters.

It would have been nigh impossible for him to renege on these relationships, in consideration of the likely repercussions upon his family and reputation. I believe that it was this dilemma, which caused him such mental anguish, the inability to remain totally loyal to John Newman, his revered friend, without harming his family in some way. John Henry Newman was eventually ordained as a Catholic priest in Rome (May, 1847), and was elected to the College of Cardinals (15 May, 1879), aged 78. Cardinal John Henry Newman chose as his semiotic emblem of office, **'Cor ad cor loquitur': 'Heart speaking to Heart',** in dialogue with the **'Loving Sacred Heart of Christ, concerning the true meaning of the Will and Word of God'.**

We may be sure that in his inner heart, Bishop Edward Pusey joined in this prayer of thanksgiving, though sad at heart that he was unable himself, to forsake all the worldly considerations involved in overcoming the constraints imposed by *'Church + State'* allegiance. It cannot be just a matter of coincidence that the same spiritual, *'tug of war',* was shattering the peace of Europe at the same time. The Franco-Prussian War (1870-1871), caused by the aggressive Prussian ambition to unite Germany, by annexing the southern peaceful States of Baden, Wurttemberg, Bavaria and Hesse, as well seizing more French Territory, led to the fall of the 2nd French Empire, which was only just recovering from the trauma and destruction wrought in the Napoleonic Wars. The ensuing *'domino effect',* set the stage for the turmoil which was to follow, cascading to abyssal depths of depravity and destruction in the two World Wars of the 20th Century.

All the Royal Houses of Europe, including Britain, became embroiled in the manipulations involved in these unfolding hideous conflicts, which unleashed *'The Beast of Belsen',* as well as *'The Angel of Death',* with the annihilative power of the atomic bomb, which may now be wielded by a *'Robotic AI'.* The wounds inflicted upon the human mind and the world environment, continue to fester and poison relationships between nations and religious factions throughout the world. How is it possible to heal these symptoms of pathogenic disorder, driven by morally infectious motives of greed and profit? The most sinister aspect of these ideologies is that they have no qualms in adopting a *'religious shield',* to justify evil practice. In this way, figures of the utmost respectability, including Royal Houses, Bankers, Financial Institutions, Lawyers especially, and Clergy, have been exposed engaging in various nefarious activities, such as fixing exchange rates (Libor); using clients' funds for gambling purposes; rocking the stock market for the purpose of making overnight profits at the expense of the small investor; raiding pension funds; paedophilia; fratricide; genocide; manipulating legal processes to hold others in servitude, etc. When Jesus Christ was counselling against the danger of acquiring riches, which is closely connected with the acquisition of power, he was asked how it was possible for anyone to enter heaven, because we are all tarnished in this respect, either by having to pay allegiance to the powers that be, or by the lure of a prosperous way of life.

His reply was unequivocal: *('Things that are impossible for man are possible for God')*: Luke 18. 26-27). The meaning of those few words is of profound significance, for I believe that Christ is telling us that *'The Creative Power of Spiritual Mind'*, by which we were brought into being, acknowledged by agnostics in various ways, such as *'molecular energy gradients, random chemical reactions'*, etc., can neither be belittled as the will of a vengeful, judgemental God: the erroneous doctrine which gave rise, for example, to *'The Spanish Inquisition'* , and continues to obscure the true meaning of *'The Will and Word of God'*, by spawning false ideologies, such as *'Scientology'*, and similar weird cults. The central concept to be kept in mind, concerning *'the wonder of our being and the creation of Life'*, as emphasized in the teaching of Jesus Christ is, *'The supreme divine power of Love, emanating from God's Holy Mind'*. Surely, it behoves us to reflect that empowerment truthfully in all we think, say and do? United in that way all wounds that divide may be healed. Therefore, there is an urgent need to check and rectify the erroneous concepts, false misleading information. as well as misinterpretations of language, which have turned so many well-intentioned people away from religious practice, by obscuring the meaning and significance of the *'The Word of God'*, as revealed by the teachings of Jesus Christ.

*The belittlement of women as inferior or second-class citizens, treated as slaves. This fallacious perception of woman, as of lesser value than man, is not only cruel and hard hearted, but stems from a denial of the 2nd Commandment, *'Love your neighbour as yourself'*, replacing *'The bond of loving care'*, bestowed by the creative *'Word of God'* to hold the *'family of man together as One'*, with the harsh bondage of possession. Instead of being hallowed as, *'The Temple of Life'*, womankind has been devalued, regarded as chattels.

Since all sensory and cognitive faculties must be reverenced as gifts of *'The Holy Spirit'*, such iniquitous deception must be considered as gravely blasphemous: *('When the time comes, the Holy Spirit will teach you what to say. Everyone who says a word against the Son of Man will be forgiven, but he who blasphemes against the Holy Spirit will not be forgiven'):* Luke: 12. 10-12.

As a result, the healing quality of empathetic care, naturally endowed in the feminine psyche, has become coarsened in many ways, so that young women -in China for example, no longer want to have children, and are even prepared to terminate the life of an unborn child. Consequently, Marriage is no longer widely respected as the holy sacrament, by which a man and woman become *'one person'*, sanctified in the words of Christ, as united by God in a bond of love *'that no man can put asunder'*. Instead, it has come to be considered as a temporal agreement, which can be set aside at will, reverted to a primitive *'animal status'*, rather than a highly cultured *'cognitive fusion of two minds'*, united by the wonderous **'sacramental grace of Love streaming from God's Creative Mind'.** That is how people are turning away from *'religious belief and practice'*, with minds conditioned to think in terms of living in *'a secular world'*, free of moral constraints. Consequently, the family unit has become devalued as the vital foundation of society relationships. That leads to the depreciation of *'Human Life'*.

Confusion which has arisen between *'Spirituality of Mind & Soul'*, and *'Religious Practice & Church attendance'*: The **'Christian way of life'**, which had spread across the Mediterranean Region, by the end of the 2nd Century AD, to as far as India in the East, and probably to W. Ireland-in the form of monastic cells, was not what has come to be understood as a Religion, defined as a congregation of Church worshippers. Rather it should be understood as the **'Assembly of the followers of Christ'**, dedicated to behaving and living according to His teaching, loved and revered by all who knew Him as, *'The Healer and Son of God'*. It was a truly *'religious' way of life, that bound all together with bonds of love and care for each other'*, based on the universal faith held by all in common. *(GK. Katholou = catholic, universal: believed by all)*.

The course of human history is littered with destructive events, associated with *'religious disputes and wars'*, arising from the deliberate misinterpretation of the very meaning of *'religion & religious'*. From time immemorial, *'religious'*, has carried the meaning (semantic message), of being bound together, expressing awareness that, *'Life, living things and the state of being alive'*, have a very special significance (semiotic opening of mind), enabling us to understand that we *'share a world together as a family home'*. This enlightenment further inspires a higher mode of appreciation of one's self and the world, united together, as kindred beings living in a mystically beneficial state of existence.

That enhanced view of Life and our surrounding world generates the spiritual qualities of *'Love and Ecological Wisdom'*, in the human mind. In this respect, the language of geographical experience may be read as a *'semiotic code'*, carrying true meaning from age to age, in the form of valid *epistemological knowledge.* Thus, we are all religious, in the sense of being bound together, responsible for each other. All who were drawn to listen to Jesus Christ teaching became aware that their minds were being opened, **to see the processes of Life and our Earth-world in an enlightened way, imbued with creative meaning and purpose, a new spirituality of outlook,** enthused to understand everything in the light of **Love and Truth.**

That is how they came to describe their assembly as united in *Faith, catholic* (belief of the whole assembly), and *orthodox* (right way of thinking), confirming that, **the observant**

human mind is an essential interactive cognitive process in the systemic dynamics with which it may be engaged. From ancient times, it has been acknowledged that, *'communication between heart and mind'* can exert a strong influence on the way we see ourselves in connection with the physical world, as well as in comparison with different perceptions held by others across the world. This may have a bonding effect, leading to the formation of mutual alliances: but, more often it appears to instigate an *ethos* of competition for material gain, culminating in conflicts and violent war.

Also, geographical experience shows that religious associations, viewed as systems of human organization are dependent upon maintaining the spiritual qualities of mind upon which they were founded, and which brought them into being. However, every system of human organization is prone to split into factions, arising from differences in opinion, and the frailty of human nature, to interpret the meaning of concepts and words according to changes in circumstances.

Thus, *'Christian assembly'*, has become fractionated, sometimes by a special focus on a certain aspect of Christ's message to His assembly: e.g., The Dominicans are noted as a highly educated *Preaching Order;* The Jesuits, as a *Teaching Order,* fixed on strict adherence to orthodox interpretation; The Franciscans, dedicated to living in poverty and sharing their lives with the *'least of Christ's brethren';* The Benedictines, devoted to the *Rule of St. Benedict,* the gentle peace-loving Abbot, who in the 6th Century, held the *'Christian assembly'* united in faith and endeavour with the simple precept, *'Laborare est Orare',* *according to the example set by Christ's life and unfolding mission,* expressed as a system of guidance and control: *('Lectio Divina').* With the collapse of the Roman Empire, that was a time of great turmoil and unrest, and fear of worse to come. The Benedictine demonstration of the healing power of *'Faith',* remains an enduring lesson to our troubled world today.

(Gk. ortho=straight + doxa=opinion: orthodox = correct way of thinking).

But undoubtedly, the most unsettling cause of the split into religious factions, arose from the close alliance forged between *Church* and State, and the embittered relationships between nations, ensuing from their competition with each other for economic supremacy and dominant power. Inevitably, the corrosive effects of double-dealing and deceitful diplomacy, has brought leading prelates into disrepute, tarnished for lending support to regimes of tyrannical rule. That is what follows, when the cohesive *'bond of love and care'*, becomes weakened and replaced by a divisive outlook of suspicion and contempt, smouldering, and ready to erupt into conflict and war. Once caught up in these coils of corruption, it is very difficult for Churchmen of good heart to escape. They become too useful as a cover of moral respectability, for the *'Machiavellian intriguers of State'*, to let them go unscathed. That is what happened to Cardinal Wolsey, when he had second thoughts about condoning psychopathic excesses, on hearing of the execution of St. Thomas More for opposing the king's evil desires. Summoned by a military escort to report to the court, he died on the way, probably assassinated.

Corrupt practice in high places paves the way to more insidious iniquity, unless checked by men of good faith. That appears to be the dilemma facing Anglican reformers, longing to restore the vitality of Christ's avowed presence in the liturgical ceremonies of the Church of England. Recalling a comment made by Sir Peter Hall, my research tutor at London University, concerning the validity of Planning Legislation associated with sound Urban development, *'**Any system of human organization based on false premises is bound to fail'**;* let us hope and pray that our Anglican brethren persist in their mission to restore *'The Assembly of Christ to England's green and pleasant land'*, not forgetting Wales,

Scotland, Ireland, as well as Cornwall, the Isle of Man and the other 5,000 smaller islands. Christ's last prayer was *('Ut omnes unum sint'):* Let all be One!

The Oratorian Bond of Love: I recall my sense of elation and expectation of good things to come on my first day at London Oratory School, like entering an *'Aladdin's Cave'*, full of intellectual treasure, but I had yet to find the really precious store of wealth, waiting to be discovered there. I could not have known there was an immensely more valuable hoard of wealth, an educational and cultural *'spirituality of mind'*, available to all disposed and ready to receive it. It must have taken two more years for me to realize fully, that my mind had been opened in an extraordinary way, like the warm glow one experiences on greeting someone admired and loved, and it dawned on me in unexpected ways. For example, when I first heard words spoken in French, it was as if I had heard and spoken these *'phonemes'* before: I recall the word *'Fresnoy'*, when my French teacher asked if any of us knew how to pronounce it. When I put up my hand without thinking and replied, he looked surprised and asked if we spoke French at home.

I found myself equally nonplussed, because I had not heard words spoken in the French language before, but many years later discovered that my Great Grandfather x 3 removed *(named Denieffe)*, had escaped from imprisonment in the Bastille during the French Revolution, probably by giving up his possessions as a bribe to the gaoler, and had settled in Ireland. Is memory a matter of reconstruction, rather than recall? Such experiences give reason to believe that the ***'Word and Will of God'*** is constantly active, educating the human mind and influencing human behaviour, and should not be regarded as just words of advice.

On another occasion, one of the priests at Brompton Oratory- I think it may have been Fr. Dale-Roberts, came into the classroom carrying one of the old plastic records, which he proceeded to play for us on an antique gramophone: it was, *'The Dream of Gerontius', John Henry Newman's poem, set to music by Edward Elgar.* He then asked if we liked it. Of course, this evoked an effusive response of appreciation from the class. In reply he beamed and said, **'Boys, we have two parts, an earthy part & a higher spiritual calling of mind: always follow what your mind is telling you. It will lead you on'.** Those few words continue to reverberate in my sense of consciousness, not always heeded, but still gathering in meaning, 85 years on. It was echoed in the hymn which we regularly sang: **'Lead kindly light, amid the encircling gloom, Lead Thou me on. The night is dark, and I am far from home, lead Thou me on'.** In the course of time, we learned that Cardinal John Henry Newman, our *'School Mentor,'* had described his spiritual journey in the steps of St. Philip Neri as, **'Out of darkness into Truth',** the religious way of life delineated in Christ's invitation to Simon Peter and his brother Andrew, the fishermen He met on the shore of Lake Galilee, who became His first disciples, **'Follow me and I will make you fishers of men'.** Scarcely literate, they could not have known that they were being called to *'open and educate the human mind',* as understood by St. Philip Neri, and with even clearer insight, by Cardinal Newman.

In that respect St. John Henry Newman, must be honoured as an outstanding Educator, with an unquenchable zest to inform and open the minds of his pupils, as shown by his comment shortly after founding the Oratory School in 1863: **('Catholics in England, from their very blindness, cannot see that they are blind. To aim then at improving the condition, the status of the catholic body, by a careful survey of their argumentative basis, of their position relatively to the philosophy and the character of the day, by giving them juster views, by enlarging and refining their minds, in one word, by education, is (in their view) more than a superfluity** *or a hobby, it is an insult. It implies that they are deficient in material points. Now from first to last, education, in this large sense of the word, has been my line'):* 'Newman 'The Heart of Holiness':Ch.9: 'A Talent for Educating':pp.113-130: Mon.Roderick Strange: St. Mary's University).* **(69).**

He is to be remembered therefore, as one of the greatest teaching minds of the 19th Century, with a pioneering interest in Human Cognitive Psychology, foreshadowing the work of Jerome Bruner, a hundred years later, who developed the profoundly important *'Theory of Education',* which continues to influence development and progress in all fields of Education, with the insight that, *'The Human Mind has the capacity to learn more from what has been learnt'.* In this context, his study of what a University could be, must be considered a pioneer work, in establishing the new science of *'Cybernetics',* as a major cognitive system of integrative regulation and control in *'Life Processes',* linking systemic concepts of *'thinking and perception',* with *'learning and communication: 'The Idea of a University':* John H.Newman *(1852).*

Bruner (1915-2016), was convinced that *'sensation and perception',* are active rather than passive processes, publishing these ideas in 1956: *('The Study of Thinking')* and developed in 1990: *('Acts of Meaning').* This train of thought resonates with the thinking of Ludvig Bertalanffy-the founder of General Systems Theory, as well as with the findings of the distinguished Neurobiologist, Paul Weiss, pioneer of Developmental Biology, with special reference to basic neural patterns of co-ordination. Bruner's conceptual understanding of *'Learning',* as a cognitive process which cannot be separated from *'Life Processes',* also reflects the perception of *'Life',* as *'a chain of informed teleonomic development',* in which *'contingent molecular systems'* appear to communicate with each other: *('We intuitively understand the operation of the biological world, including of course, all human activity, through Life's teleonomic character. In the non-living world, by comparison, understanding and prediction are achieved on the basis of quite different principles'): 'What is Life': pp.9-11: Addy Pross, Professor of Organic Biological Chemistry).* **(70).**

But, without the Human mind there could be no concept of difference between biological and non-living systems. Without the capacity of human consciousness to be cognitively aware, there could be no perception of organization and regulatory order in systems. There would be no recognition of systemic functions at all. Indeed, it can only be through the mysterious endowment of *'principles of logical thinking'*, in our minds, that we are able to contemplate whether there can be any fixed laws, or *'constants in Qualitative Electrodynamic Systems'*.

Language can be interpreted to confuse the meaning of Christ's message. Cardinal Newman believed that all we come to know *(Knowledge),* understand *(Semantic interpretation),* and behave in what we do, can shape what happens in the world *('Semiotic wisdom): ('A cultivated intellect, because it is a good in itself, brings with it a power of grace, to every work and occupation which it undertakes, and enables us to be more useful, and to a greater number'): 'The idea of a University': John Henry Newman: Quoted in 'Newman: The Heart of Holiness': Mon Roderick Strange: St. Mary's Un.)*

In this didactic manner, John Newman is telling us that the world, in which we have come into existence with all life, is not to be dismissed as a *'secular state of being'*, outside of ourselves and devoid of rules. Nor must we be drawn into the materialistic trap of devoting our life efforts to the gain of wealth and power. Neither must we be deluded into thinking of our purpose in life and our geographically benign home-world in a deterministic way. Rather, it is essential to understand that all that we are, see, and know must be esteemed as *'probabilities of a spiritual dimension',* coming into existence as true knowledge, through the mystical cognitive processes of our human mind. In this respect, he must be regarded as a scientist, well ahead of his time, aware that there is **'A principle of Universal Order, by which all knowledge has a unifying effect',** in opening the human mind to see Life and the parameters of the Geophysical Environment, as a web of interacting systems.

('The world is in a ceaseless process of change:

the entire evolution of science would suggest that the best grammar for thinking about the world is that of change, not of permanence. Not of being, but of becoming. We can think of of it as being made up of things; of substances; of entities; of something that is. Or we can think of it as made up of events; of happenings; of processes; of something that occurs: something that does not last, and that undergoes continual transformation, that is not permanent in time The destruction of the notion of time in fundamental Physics is the crumbling of the first of these 2 perspectives, but not of the second; it is the realization of the ubiquity of impermanence, not of stasis in a motionless time. Thinking of the world as a collection of events, of processes, is the way that allows us to better grasp, comprehend and describe it. The world is not a collection of things (material), it is a collection of events. Things persist in time; events have a limited duration: a stone is a prototypical thing; we can ask where it will be tomorrow. Conversely, a kiss is an 'event': no sense to ask where it will be tomorrow. The world is made up of networks of kisses, not of stones'):'The Order of Time': Ch.6:'The World is made of events not things':pp.85-92: Carlo Rovelli: Published 2018). **(71).**

This view of the world, heralds a new science of **'Philosophical Physics',** replacing the worn out *'Cartesian'* model of Life and the physical world as mechanisms, which in turn fostered an outlook of materialism and godliness, paving the way to the dead-end quagmire of hatred and conflict, which boiled into the horrifying events of total war in the 20th Century. Why were the sciences jerked out of complacency, and rocked to their crumbling foundations, just before the outbreak of WW2, by the revelations of '*Quantum Theory*', that the molecular systems underpinning conceptual thinking of ourselves and world events as '*real and fixed*', could no longer be upheld? Was that a wake-up call to our human mind, alerting us to what might befall, as a consequence of the apathetic state of materialistic pessimism sweeping through the world, with nations stockpiling munitions and lethal weapons in preparation for all-out war? My '*Oratorian*' education inclines me to believe that Cardinal Newman may have been enlightened by the same sense of *Scientific*, as well as *Theological* sense of foreboding, that '*world order*' was about to fall apart yet again, as the embittered relationship between France and Germany began to flare into the Franco=Prussian War (1870-71). The very word, '*hamburger*', reminds us of the queues of people crowding the docks in Hamburg, fleeing to a safe haven in the '*New World of N. America*', desperate to escape from the pogroms which were taking place at the same time, mainly in Russia and Germany in fear of worse to come, pausing only to grab a hasty sandwich. The same pessimistic view of the world could have been expressed by John Henry Newman in the 1850's, except that his perception was coloured by an insight of enlightened discernment, that enabled him to see that the world was not inevitably running down to a state of disorder; **condemned to ultimate destruction and extinction, like *"Shiva's flailing swords"*, leading to a state of increasing '*Entropy*' and oblivion.**

('We are glimpsing something about the mystery of time. We can see the world without time: we can perceive with the

mind's eye The profound structure of the world, where time as we know it no longer exists. And we begin to see that we are time. We are this space, this clearing opened by the traces of memory inside the connections between our neurons. We are memory. We are nostalgia. We are longing for a future that will not come. The clearing that is opened up in this way, by memory and anticipation, is time: a source of anger sometimes, but in the end a tremendous gift. A precious miracle that the infinite play of combinations has unlocked for us, allowing us to exist. We may smile now. We can go back to savouring the clear intensity of every fleeting and cherished moment of the brief circle of our existence'):'The Order of Time': Ch.13:'The Sources of Time': pp.167-175: Carl Rovelli). **(72).**

By stark contrast, St. John Henry Newman's perception of Life and the World stemmed from an outlook of *'Theological Faith',* as well as *'Theosophical Wisdom'.* That optimistic view depends on whether the Human Mind remains spiritually receptive, and not shut down by false information, such as, *'we live in a secular world, where everything is up for grabs';'if you don't look after yourself, no one else will';* *'religion has done more harm than good',* etc. In that debate, concerning *'Unity of knowledge and the emergence of what may be perceived as the **reality of existence',*** Professor Roger Penrose, distinguished polymath and metaphysician, raises a searching question: **('How is it that the mere counterfactual probability of something happening- a thing that might happen, can have a decisive influence upon what actually happens?'.** He draws attention to the parallel between, **'collective (coherent) quantum effects observed to be operating in biological structures (human beings), as well as at the scale of molecular systems (cells, neurons)':** *'Shadows of the Mind: A Search for the Missing Science of Consciousness': Sir Roger Penrose).* **(73).**

173

Thus, our human mind appears to be endowed with the capacity to act like **a quantum (qualitative) processor, as a system of discernment and control, capable of comprehending the quantum (qualitative) effects, of probabilities emerging and coming into existence as realities of geographical experience.** Was that not the same message conveyed by the ancient Greek strictures, *'Know thyself'*, and *'All things in moderation'!* The same message was spelt out in words of unmistakable meaning and clarity by Jesus Christ to His Apostles, to be understood and acted upon by every member of His Christian Assembly: *('Accept and submit to the word which has been planted in you and can save your souls. <u>But you must do what the word tells you, and not just listen to it and deceive yourselves.</u> Pure unspoilt religion, in the eyes of God our Father is this: coming to the help of orphans when they need it and keeping yourself uncontaminated by the world': 'Take the case of someone who has never done a single good act, but claims that he has faith. Will that faith save him? If one of his brothers or sisters is in need of clothes, and has not enough food to live on, would you say, 'I wish you well; keep yourself warm and eat plenty', without giving them the necessities of life, then what good is that? Faith is like that: <u>if good works do not go with it, it is quite dead</u>': 'Wherever you find jealousy and ambition, you find disharmony, and wicked things of every kind being done, whereas the wisdom that comes down from above is essentially something pure: it also makes for peace, and is kindly and considerate; it is full of compassion and love and shows itself by doing good; nor is there any trace of partiality or hypocrisy in it. Peacemakers, when they work for peace, sow the seeds which will bear fruit in holiness. <u>Where do these wars and battles between yourselves start? Isn't it precisely in the desires fighting inside you</u>?'):* Letter of St. James: 1.17-27:3.16-43).*

John Henry Newman, as a dedicated Anglican priest, clearly took these words to heart, contemplating the Church of England falling into disrepute, exposed as a sham form of religious observance. It caused him dismay to see what had been regarded as, ' *celebration of Christ's message in the Holy Mass,* reduced to a hypocritical theatrical performance, paying lip-service to the *'Words of Jesus Christ'*, whilst prepared to ignore *'Christ's Message, Love your neighbour as yourself'*; which means in practice that our main purpose as a good rule of life should be understood as, **giving rather than taking; of serving rather than dominating; of educating rather than suppressing; and of healing rather than inflicting wounds.** It appears that he became convinced that all knowledge and experience tells us that the *'Word and Will of God'*, must be regarded as a creative unifying power constantly working to perfect and bring forth a loving world ethos for the good of all.

In this respect, he foresaw the *rapprochement* of scientific revelations with theological doctrine, which we see taking place today, with the realization that a materialistic view of Life and the world is illusory and can no longer be upheld. The implications are profound. What we experience as *'realities of geographical existence'*, (Life Processes, linked with Cognitive Faculties interacting with Geophysical Systems), we now know as the emergence of the highest probabilities, shaped by all that has gone before, including historical influences from time and minds long ago, as well as by all we think and do. Therefore, we learn that good begets good or better; hate promotes hostile relationships, conflict and war; so does the uncontrolled lust for wealth and domination: whereas a kindly generous outlook commands a loving response, which fosters an outlook of optimism and hope towards our fellow men (and women), and engenders a spirit of *Faith, in communication with the Mind of our Creator God.*

175

Thus, I believe that John Henry Newman's great leap of faith, was his unswerving decision to eschew all erroneous influences that, in his view had led to the dissolution of *'The Christian Assembly'*, by obscuring the meaning of Christ's message. In that way, he was drawn to the *'ethos of the Oratorian Brotherhood'*, (Congegation of the Oratory), founded by St. Philip Neri, and revered as *'The Apostle of Rome'*, who followed Christ's didactic words to the letter, by working unremittingly day and night with loving care, amongst the sick, poor and dying to alleviate their sufferings. At that time in the 16th Century, the streets of Paris were littered with the corpses of children, reduced by starvation to skeletal figures of skin and bone, just as in Ireland during the 18th-19th Century, where you may still see the hillside graveyards *('Gorta')*, with graves marked by thousands of triangular stones, small ones for children, and slightly larger for adults. That, we may learn, is the way *'God's Will and Words'* are constantly vibrant and working, by educating the human mind and influencing human behaviour: thereby shaping the nature of events which next come into existence.*(Gk. ethos=nature, disposition influencing behaviour: didactikos from didaskein=to teach, educating by opening the human mind through teaching and learning).*

Counteractive influences: When horrible things happen, such as *'environmental hazards associated with climate change'*, and *'Covid 19 Pandemic'*, that is when those blinkered by *'secularist materialism'* join with atheists grumbling, *'a benevolent God wouldn't let such things occur'*; forgetting that we are part of the equation shaping and influencing the nature of future events and geographical experience. Christ's words told us in advance, to be on our guard against such counteractive influences, showing *'neither fear of God nor respect for man': ('Now, will not God see justice done to His chosen who cry to Him day and night, even when He delays to help them? I promise you, he will* *see justice done to them and done speedily. **But when the Son of Man comes, will He find any faith on earth?'**): Luke 18. 1-8).*

We may also learn from these words, that the whole *'Fabric of Creation'*, must be unfolding according to a *'principle of Universal Order'*, from the beginning to the end, *'alpha to omega'*, just as the great Theologian, St.Thomas Aquinas told us in the 12ᵗʰ Century. *(Quinque Viae: 5 Proofs of God's Existence).* The Book of Wisdom always has some useful advice to offer, to help us see ourselves as we are, as well as to know what may result as a consequence of what we choose to believe and behave: *('Does the pot say to its maker, "Thou hast no skill").* What should be expected to happen, shortly after we have caused whole Islands and Mountains to disintegrate and vanish before our very eyes, from testing nuclear weapons? Do we find pleasure in unleashing such destructive power? We have learned that hostile acts and hateful behaviour always lead to violent conflict and destructive war, whilst also knowing that an ethos of loving care will invariably evoke a response of harmonious peace and mutual trust. The choice is ours, we have the option to bring down *'Nemesis'* upon ourselves, or to **bring forth a better and happier world order, for the good and well-being of all.** That encapsulates the *'Principle of Universal Order'*, underpinning Life processes linked with the physical systems of our earth-world home. *(Gk.nemesis=righteous indignation: from nemein=deal out what is due, deserved: ethos=nature).*

It would a fundamental error to assume that our **Creator** is a judgemental Deity, ready to dole out punishment for misdemeanours of a criminal nature and acts of wicked behaviour. The truth is, that the task of judging human behaviour has been delegated as an obligation of orderly organization within the cognitive faculties endowed in, *'the human heart and mind'*. There can be no doubt that we are responsible, for what we do. The astonishing scientific revelations arising from *'QED: Quantum Theory'*, associated with the informative *'molecular chains of instruction'* involved, have opened our understanding, in this respect. For example, how can every one of the 30 to 40 trillion cells in my human body know more about me and how I am made, than I can ever learn during my lifetime? The sciences are converging with theological teaching, to explore this exciting field of knowledge, unfolding as the new science of **'Cybernetics'**, concerned with the study of the human mind, as an integrative system of control. *('**Molecular Forces must have been 'fine-tuned' to make Life on Earth possible'):* 'Just 6 numbers': Astronomer Royal, Lord Martin Rees).* **(74)**.

Consciousness must be held as precious as life itself, recognized as a fundamental system of comprehension, necessary to understand the discoveries of scientific endeavour as chains of interacting connections, knowing that full understanding of the network of complex relationships (of which we are a part), as ordered systems of organization, is beyond our mental capacity. It would be entirely fallacious and misleading to perceive our Earth-world as a place, redolent of hate and hardship, which so many millions striving to survive in conditions of dire poverty, find it to be. In fact, all the scientific knowledge, gleaned from astronomical observations, as well as studies of Earth's interacting geophysical systems, show our home world to be remarkably prepared to support *'Life Processes'*, **just right, gravity bound, and equipped with astonishing intricately ordered support systems:** *('Life depends on a wonderfully ordered and intricate system of both simple and complex molecules':* 'The Goldilocks' Enigma': Paul Davies).*

You may think it natural, for those who are fortunate to be located in a productive region of the world, with a good style of life, well-fed and educated, to appreciate the beauty of our home-world in this way, inspired to help the countless number of their fellowmen condemned to a life of misery, lacking even the barest necessities, such as clean water. 2 billion people, a quarter of the world's population are estimated to be drinking filthy water, contaminated with disease and amoeba, at the same time as $2 billion was wasted in sending a space probe to investigate an asteroid. Vast resources are being squandered in the development of *'space travel pleasure tourism',* for the ultra-rich, whilst little girls as young as 6 have to walk many miles a day, over rough stony tracks, in blistering tropical heat, to carry back a heavy load of dirty water from muddy water-holes, riddled with disease bearing bacteria. In May, 2020, it was reported that one of the richest billionaires had joined the queue of 2 million children in the Uk, having to go hungry to school, and in need of food vouchers. The excuse for this display of selfish temerity, in seeking Government aid, was that *'Covid 19 Pandemic'* had interfered with the multimillion-dollar development of a tourist space carrier, and caused extra expense. Hateful and uncaring human behaviour will inevitably bring forth a worsening world order, with rumbling resentment and discontent, leading to conflicts and hostile relationships, which if unchecked, may fester and erupt into more traumatic events, such as those which occurred during the French Revolution, with the continuing corrosive effect of souring the human mind. It would be well for those who have become addicted to an insatiable desire for wealth and power, to be reminded that whatever affects the lives of other people in the world, will recoil upon themselves as a judgement, shaping the nature of the ensuing state of their existence, as they deserve.

('There was a rich man who used to dress and feast magnificently every day, and at his gate a poor man called Lazarus, covered with sores, who longed to fill himself with the scraps that fell from he rich man's table. Now, the poor

179

man died and was carried away to the bosom of Abraham. The rich man also died and was buried. In torment in Hades, he saw Lazarus a long way off with Abraham. "Father Abraham, pity me and send Lazarus to dip his finger in water to cool my tongue, for I am in agony in these flames." Abraham replied, "My son, during your life good things came your way, just as bad things came the way of Lazarus. Then the rich man replied, "Father, I have 5 brothers, I beg you, "Send Lazarus to warn them, so they do not come to this place of torment too". Abraham replied, "If they will not listen either to Moses, or the prophets, they will not be convinced, even if someone should rise from the dead"): Luke 16. 19-31. Our Lord, Jesus Christ, did not condemn anyone either, even when He was dying in agony on the Cross; (*'Father, forgive them, for they do not know what they do!'*). Neither did He condemn the sins of men, in what He said, when he appeared to them after *'The Resurrection'*, which we continue to celebrate in the **Catholic Liturgical practice of holy Mass,** in recognition of the continual presence of Christ amongst His assembly, as He requested at *'The Last Supper'*: *'Do this in memory of Me!'*.

All, who claim to be Abraham's children would be well advised to keep their Patriarch's words close to their hearts, in view of their hateful behaviour towards each other, and the resulting hideous loss of life and environmental destruction it has caused. They have even found it convenient to misinterpret the very meaning of their own family names, to find reason for killing each other: *Arab, (Gk.Araps=member of Semitic people);* **Semites,** *(Gk. Sem (Shem)= son of Noah), Semitic People include Jews. Arabs, Babylonians & Phoenicians;* **Hamites,** *(Gk. Ham (Cham), 2nd son of Noah);* **Muslim,** *(Arabic, aslama= submit to the* **'Will of God';** **Islam,** *The Muslim Religious Assembly founded by Muhammad, meaning the people who have submitted to* **Allah,** *(Arabic: al-ilah= The God).* Thus, the very meaning of the '**Word and Will of God',** may be subverted to mean the very opposite of the true semantic interpretation. Similarly, the semiotic guidance of **God's Words to all people, ('Love one another as you love yourselves'!),** has been obscured. The same lesson applies to the antagonistic behaviour of Christian factions between each other, such as the '*Orange Order'* triumphal marches, which take place in N. Ireland, in commemoration of bloody battles and associated atrocities, such as the burning of a church in Drogheda with hundreds of people herded inside, in 1649. That is the way that '*The Words of God',* have been continually obscured and twisted, to suit various nefarious purposes. However, we may take heart from the inspiring words of Jesus Christ: *('You are Peter, and upon this rock I will build my Church. And the gates of the underworld can never hold out against it"):* Matt.16. 13-20. Peter, inspired with confidence and absolute *'faith' in the word of Christ,* was further graced with an opening of mind to understand sufferings, such as we are witnessing everywhere in the world today, including the effects of evils we have brought upon ourselves and inflicted upon each other, as a **'cognitive immunological response', of guidance and control, to bring us back to our senses as members of the "Assembly of God's People":** *('Be calm but vigilant, because your enemy the devil is prowling around like a roaring lion, looking for someone to devour. Stand up to him, strong in faith and in the knowledge that your brothers all over the world are suffering the same things. The God of all grace who called you to eternal glory in Christ will see that all is well again: He will confirm, strengthen and support you. His power lasts for ever and ever'):* Peter 1. 3-11.

We are reminded time and time again, how the *'Word of God'* works continually in the heart and mind of man (and woman), for the well-being of all. The world cannot be considered to have come into existence by random chance. Neither can the mysterious emergence of Life and living things be dismissed as an aimless and haphazard event, as some shallow thinking scientists, bewildered by the mind-boggling network of contingency interactions linking *Life processes (including the Faculties of Consciousness)* with (*'Geophysical environmental parameters),* have been prone to suggest. Studies of *'QUED (Qualitative Electro-dynamics),'The Human Genome' (the complete set of genes & genetic material),* contained in every one of the 30 to 40 trillion cells of a human body (30,000,000,000,000,000), confirm that all we know and experience as *'realities of* Geographical **existence, has been "fine-tuned from the beginning of Creation"):** *'Just 6 Numbers': Lord Martin Rees, Astronomer Royal).* Even the chains of *'informative molecular communication',* required to convey the precise parameters of the previous step, in order to ensure the exact specific properties necessary to continue millions of ongoing contingent reactions, in order to each the appointed purpose, proclaim **a Universal Principle of Order uniting all that has been brought into being,** which cannot be gainsaid or frustrated in any way. One word missing out of billions, and the existence of Life, the world, you and me, would not be possible.

Christ's words told us that. How could the human mind become so obtuse, as to let such vital knowledge go in one ear and out of the other? That is what happens when the saving grace of *'Love in the human heart and psyche',* is replaced by the hateful influence of self-seeking vice, in pursuit of wealth and power. In this way, as we should have learned from the history of past traumatic events, human life becomes devalued, and human beings who stand in the way, disposable.

St. John Henry Newman's great achievement: It is my firm conviction, based on my Oratorian education, that Cardinal Newman's remarkable contribution to the re-establishment of order, and the restoration of peaceful loving relationships in our war-torn turbulent and floundering world, was his insight that harping on past conflicts and sufferings, has been equally divisive as the idolatrous worship of wealth and power, driven by a determination to dominate and rule the world. He was able to see how false conceits of this kind become pervasive and destructive influences, prolonging hateful festering conflicts of the past. His educative mind enabled him to see through the web of chicanery, involving collaboration of Church and State, which had reduced Catholic liturgical worship, centred on belief in the presence of *Christ,* to the enactment of a theatrical performance, in the Church of England, set up by Henry VIII in 1533, to satisfy his own psychotic desires.

Thus, I believe, like St. Francis of Assisi, Cardinal Newman was being called to restore '*the Christian Assembly*'. It would have been tempting to let bygones be bygones, rather than incur the risk of antagonizing powerful men of State and Church, who were enriching themselves handsomely at that time in Victorian England, from the spoils of Empire building. He realized that the '*blasphemous enormity of desecration*', which had torn the '*Christian heart*' out of liturgical worship in the so-called '*Church of England*', could not be glossed over. If condoned and left unchecked. he could see that its cankerous effects would continue to spread like a pathogenic spiritual disease, denigrating religious practice, as well as the very '*Word of God*'.

Consider, for example, the hideous events following '*The Dissolution of the Monasteries' (1536-1541),* after Henry VIII had declared himself as '*Supreme Head under Christ of the Church of England', in 1533:-* Over 300 recusants were summarily condemned to an agonizing death, by burning at the stake (1533-34): In 1535, 6 Priests were hanged, drawn and quartered, as public spectacles of entertainment, and 10 monks were deliberately starved to death in Newgate Prison:

In 1537, 216 were executed, including 38 monks. Within a few years, monasteries were desecrated and left in ruins throughout the length and breadth of England, Scotland, and Wales.

The ruins of magnificent church buildings can still be seen. Those who rose in revolt, in Devon and Yorkshire, were suppressed and put down with merciless cruelty. Following the harrying of Yorkshire, it was recorded that corpses were still where they fell in the hedgerows, decades later. Henry VIII evidently derived psychopathic delight from afflicting hideous torture upon anyone who dared oppose him. When the wife of Robert Aske, one of the leaders of the Yorkshire rebels was condemned to death, went personally to plead that his sentence to death by being hanged, drawn, and quartered, could be commuted to hanging. Henry agreed, on the grounds of the *'chivalric code of honour', that a woman expecting a child', could request lenient treatment for the father'*. But, when the sentence was carried out, Robert Aske was bound by a chain and hoisted to be hung on the flagstaff of the tower in York, where he hung moaning in extreme agony for 3 weeks, before he died. Margaret Clitheroe of York was condemned to death for sheltering a priest, suffering a martyr's death by being pressed between 2 doors by adding heavy weights. The full record of human suffering, wrought under the reign of the Tudor dynasty can never be known, but the corrosive effects still continue as a legacy of evil, to sour social relationships and emerging events. Opposing the powers that be, at that time in Tudor England, would have invited a painful death. It would have been no less risky in Victorian England to court the vengeful wrath of the money-making ruling elite. In this respect Cardinal Newman shared the anguish of his fellow Anglican Theological Reformers, priests of good faith like his friend Edward Pusey, who longed to revive the feebly beating **'Christian heart',** to its former hallowed state of vitality, in the liturgical worship of the church, profaned by Henry VIII, when even the altar tables where Holy Mass had been celebrated for centuries were ordered to be set in the floor, to be trampled upon, as a symbolic ritual of desecration.

Cardinal Newman described his travail of *'distressed spirituality of Mind'*, at that time as, *'from darkness to light'*, later expressed in his poetic hymn, *'Lead kindly light'*, our favourite hymn which we sang regularly during my time in London Oratory School:- *'Lead kindly light, amid the encircling gloom, lead Thou me on; The night is dark and I am far from home, lead Thou me on; Keep Thou my feet, I do not ask to see the distant scene; One step enough for me. I was not ever thus, nor prayed that Thou should lead me on; I loved to choose and see my path; but now lead Thou me on; I loved the garish day, and spite of fears, pride ruled my will; remember not past years. So long Thy power hath blest me, sure it will lead me on; o'er moor and fen, o'er crag and torrent till the night is gone, And with the morn those angel faces smile, which I have loved long since, and lost awhile.'* John Henry Newman (1801-1890): Music by C. H. Purdy (1799-1885).

With the advantage of hindsight, looking back at my learning experiences at the Oratory, I now see that my mind was being opened to contemplate the ups and downs of geographical existence, as an unfolding web of interactions influencing human behaviour, with the same insight that led John Henry Newman (Anglican Priest of good faith), to convert back as a Catholic Priest of the *'English catholic assembly'*. No doubt, as a scholarly Theologian, he would have been influenced by St. Paul's cautionary words, concerning *'The true teacher and the false teacher':* (*'This is what you are to teach them to believe and persuade them to do. Anyone who teaches anything different, and does not keep to the sound teaching, which is that of Our Lord Jesus Christ, the doctrine which is in accord with true religion* (religious practice)- *is simply ignorant and must be full of conceit – with a craze for*

185

questioning everything and arguing about words. All that can come of this is jealousy, contention, abuse and wicked mistrust of one another; and unending disputes by people who are neither rational nor informed <u>and imagine that religion is a way of making a profit.</u> Religion, of course does bring large profits, but only to those who are content with what they have. We brought nothing into the world, and we can take nothing out of it; but as long as we have food and clothing, let us be content with that. People who long to be rich are a prey to temptation; they get trapped into all sorts of foolish and dangerous ambitions, which eventually plunge them into ruin and destruction. <u>The love of money is the root of all evils, and there are some who, pursuing it have wandered away from the Faith, and so given their souls any number of fatal wounds'):</u> St. Paul: 1 Timothy: 6. 3 – 10.

When St Paul began his missionary journeys (40-50 AD), it is evident from his letters and sermons, as recorded in *'The New Testament',* that his view of Life and the world and of himself had undergone a miraculous change. No longer the merciless zealot, willing to kill those he considered to be blasphemers, and offensive to the God of Israel, as if it were his duty to carry out the wrathful judgement of God, but an altogether kinder and more caring reformer. We may learn a lesson of profound significance from this abrupt change of heart, in an unusually gifted highly intelligent and well-informed mind, well-versed in the Hebrew Scriptures, as well as the interpretation of *'The Words of the God of Israel in the Mosaic Law'.* **Note:-** God, our Creator is always loving, like a father to his children: Always forgiving our sins and offensive behaviour; we have only to recognize the faults in ourselves, confess our iniquity and to ask for pardon: God does not punish, but wishes to cure all the diseases of mind and body we have brought upon ourselves, but we must listen to His *'Words and understand their meaning, and not just pay lip-service to suit our own selfish desires'.*

The Will of God is being inexorably carried out, not by *'servants and armies of a material world, but by the spiritual empowerment of a "Principle of Universal Order", woven in the fabric of the tapestry of Creation.* The word *'fear'*, in this context does not ring true, and may be an inaccurate translation of the Hebrew word for *'revere'*, or *'respect'*, which makes more sense. *(Hebrew: YHWH= personal name of God in Old Testament: Scriptures: adonay=my Lord: elohim=name used in Aramaic, addressing God).*

However, St. Paul could not have been aware of all that happened in Jerusalem, after the Crucifixion of Our Lord Jesus Christ, and the events after His Resurrection, with St. Peter's address to the people of Israel, which together with *'The Acts of the Apostles'*, called **Christ's'** followers together, to form the Christian Community in Jerusalem. This momentum continued, to establish the Community in Antioch, among their, *'Hellenistic Greek speaking brethren'*, by about 30-40 AD. Saul, as he was then known by his Jewish name, had taken part in the stoning of St. Stephen, about 35 years after the Crucifixion and Resurrection of Christ, and was on his way to Damascus, determined to put a stop to the spread of *'Christianity'*, believing it to be a heretical sect, when he was struck blind by a bolt of lightning. At that moment he experienced a vision of Christ calling, **'Saul, Saul, why do you persecute me?'.** *(Acts of the Apostles: 9. 1-39).*

The electrifying effect of this encounter with the **'suffering Heart and Mind of Christ'**, transformed the nature of his cognitive sense of being, changing completely the way he perceived and understood both himself and the unfolding events in the world around him. That must have been precisely the same searing effect of enlightenment he had heard, echoed in St. Stephen's rapturous exclamation of elation, *'I can see heaven thrown open, and the Son of Man standing at the right hand of God'.(Acts:7.1-60).*

Place yourself in his shoes at that moment of his change of heart, when the hate- darkened perception dominating his psyche evaporated in a trice, causing a complete *'volte face'*, in his outlook. The corrosive sense of repugnant detestation for his brethren, whom he had judged to be at fault, was replaced by a higher cognitive sense of understanding care, with the *'autocatalytic'* effect of knowing what he was being called upon to do. As a result, he became inspired to undertake the incredibly onerous task of evangelization, in his 4 missionary journeys, (45-60 AD), with a burning desire to strengthen the ties of brotherhood between the scattered Christian communities across the Mediterranean region, stretching from India to Spain and beyond: one of the greatest feats of geographical field work ever undertaken.

Fig.9: Map of the Missionary Journeys of St. Paul: Epic feat of geographical fieldwork: Four arduous Journeys, lasting from 37 to 67 AD: A great act of submission and apostleship, to unite the *'Christian assembly of Christ's people as one ecologically-minded spiritual loving family'.*

(Gk.pharisaios from Aramaic: prisayya from Hebrew: parus=one who observed strictly the tradition of the Jewish Elders & the written Mosaic Law: pharisees).

(Gk.haeresis from haireomai from hairein= to take, make one's own= heretical cult).

(Gk. euaggelizesthai: to proclaim the 'Word of God', explained by Jesus Christ, proclaim the Gospel).

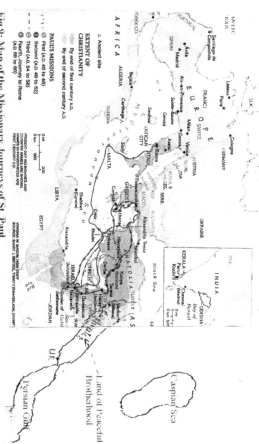

Fig.9: Map of the Missionary Journeys of St. Paul

Note: The Christian assembly, established by St. Thomas at Kerala was spreading across India, and had taken root in Madras, Orissa and Gujarat by the first century A.D. The christian message was transmitted across the Roman Empire, especially by means of Greek linguistic scholarship and the professional expertise of Greek Teachers and Medes accompanying the Legions and Administrators. Also, it was carried far and wide by way of monkish cells and hermitages, reaching as far as W. Ireland and Iran, where it resonated with the religious faith of Zoroastrianism, which still persists.

189

Unity of Science, Theology and Geographical Experience

The unifying empowerment of Christ's healing love: Saul, a man gifted with unusually high intelligence, as well as the linguistic ability to communicate with all the people of different religious persuasions, and speaking different languages, across the known world at that time, had received the healing grace which enabled him to see the interactions of the living world through the loving compassionate eyes of Our Lord Jesus Christ. Although he considered himself as the last and least of Christ's Apostles, he is revered for his charismatic role in strengthening the bonds between the widespread nascent Christian communities, growing in the old city states established by the Phoenicians across the Mediterranean region, such as Corinth, Tyre, Sidon, Byblos, Carthage and Cypru (Cyprus).

The Phoenicians originated as an ancient Judaic civilization, founded by *'Sem, son of Noah'*, in Levant (E. Mediterranean), centred in Lebanon, and stretching along Syria to Gaza in Palestine. They became a prosperous trading nation of seagoing merchants and explorers and developed a flourishing civilization which reached its peak of influence between 1100 and 200 BC, noted especially for their advanced skills in shipbuilding, navigation, industry (including smelting and metallurgy), and organized government. Therefore, they exerted a significant influence of ordered human organization, by exchanging cultural relationships, developmental ideas and valid knowledge, between Greece, Egypt and Mesopotamia, also thriving as well-established civilized states. For example, Cyprus owes its name to the mineral deposits of copper ore, discovered and worked by the Phoenicians, who developed the metallurgical techniques to produce bronze, a rust free and hard metal, superior to iron.

The Phoenician system of communication along their commercial trading network continues to influence systems of world communication today, as an efficient 'code of language', based on the Phoenician legacy of the world's oldest alphabet. We are reminded that the 'cognitive faculty of language decryption', is as mysterious an endowment in the Human mind, as Life itself, and that we may discern higher cognitive processes of understanding and communicative knowledge, in the form of complex codes such as Mathematics and Music. Possibly derived from the ancient Egyptian system of communication by means of hieroglyphic symbols, the Phoenician alphabet was used to write and communicate in the early N. Western Semitic languages (early Phoenician, Moabite, Ammonite, Edomite & Hebrew), by Phoenician merchants trading across the Mediterranean and beyond, reaching as far as China, during the 3rd millennium BC.

Success leads to prosperity, gaining impetus through propitious geographical experience', so the Phoenician language code was adopted and modified in translation, by many cultures, including the developing civilizations of the Greek and Roman Empires. In that way, the now extinct Semitic language, spoken 5,000 years ago, in ancient Mesopotamia, (Akkad, Assyria, Isin, Larsa, Babylonia), which was used to write the also extinct *'Sumerian script-written in cuneiform'*, became the *'lingua franca'* in the developing civilizations of the Mesopotamian Region (The Land between the Rivers Euphrates and Tigris.

(Gk. hierogluphicos; from hiero=holy, sacred+gluphe= carving: written in stone, set meaning: Mesopotamia, from mesos=middle +potamos=river: =Land between rivers: Euphrates from euos= well, good + phrater =clansmen, brother='Land of good brotherhood, of mutual benefit).

Mystery of The Great Flood: Gilgamesh Tablet: Arrival of Sumerian People: We now know, from advances in our ability to read the petrological evidence, contained in the formation of the rocks, that the Mediterranean basin and sea, is a remnant of a once vast ocean, *'Tephys'*, which separated the great land masses of *'Laurencia'* (N. America+Europe),& *'Gondwanaland('S.America+Africa+Aus* tralia), from the Palaeozoic (550 million years ago), to the Jurassic Era (180 m.y.a). We know also that the *'Pillars of Hercules'* (the Gibraltar gap*)*, has closed and opened several times in the course of geological events, caused by the movement of the 2 land masses, in collision with each other. As a result, the Mediterranean Sea, has dried up and been flooded again, several times. Therefore, the record of *'The Great Flood & Noah's Ark'*, contained in ancient Hebrew Scriptures and the Old Testament of the Bible, must be understood as, **a reconstructed memory of natural events which have exerted a significant impact and influence in shaping the emergence of systems of human organization, closely associated with developing patterns of social relationship, during the last 5 to 10 thousand years of human history.**

In this context, it must be kept in mind that a thousand years in human reckoning is but an instant in the geological timeline. The philosophical implication is that the concept of time must be considered as an expedient illusion of the human mind, striving to make sense of the wonder of **Life and geographical existence.** *('The incomprehensible thing about the Universe is that we can comprehend it'): Albert Einstein:* & *Blaise Pascal: Mathematician of 'Probability Theory.* Therefore, we have to consider whether the account of a *'calamitous great flood'*, as recorded in Hebraic folklore and the Bible, may be in fact an allegorical mythical remembrance of a disastrous recurring event that has destroyed previous civilizations: (cf. **Noah Flood:** dated C 2458 BC); also in the

Sumerian poems (*'Epic of Gilgamish':* dated C 2700 BC: Tablet written in Akkadian Cuneiform). *(Gk. Hebraikos from Aramaic ibray= hebrew speaking).*

Sumerian Akkadians: Who were these people, speaking an ancient *Hebrew Language,* who mysteriously arrived in a fleet of ocean-going ships, entering the Persian Gulf during the 4th to 5th millennium BC, to settle in the fertile valleys of the rivers Euphrates and Tigris? Undoubtedly it must rank among the greatest migrations of all time, survivors of a traumatic environmental hazard, such as a major flood caused by a dramatic rise in sea-level, or a prolonged Pluvius episode of very heavy rainfall. Replete with all the attributes of a highly advanced civilization, the remnants of a sophisticated nation shattered by a disastrous event, arrived in a fleet of ocean-going ships, to settle in the fertile valleys, which became known as *'The Land of Mutual Brotherhood'.* Where had they come from? Could it have been *'The Lost Nation of Atlantis',* or *'Lyonesse',* vaguely recorded in the hazy mythological memory of *'Kernow'* (Cornwall)?

We may be reminded, that everything we know, perceive, and come to believe, emerges from events and influences - including *environmental conditions+ systems of human organization+ human behaviour,* previously prevailing. But, memory or mental reconstruction of past events, may also be clouded, obfuscated or misunderstood, by nuances of linguistic mis-translation from one language to another: e.g., Akkadian (ancient Hebrew)-Aramaic idiom- Phoenician trading *'lingua franca'-* Greek-Latin- *Indo-European* languages; as well as deliberate misinterpretation for some nefarious purpose. In this context, the events associated with human behaviour arising from belief and attitudes, recorded in the sacred books of the *Abramic Religions (Judaen-Christian-Muslim),* based upon **submission to the *'Word of God',*** can no longer be regarded as sacrosanct.

They contain many erroneous concepts which continue to inflame hostile relations and bitter conflicts, with the divisive effect of fostering antagonistic and hateful factions. Thus, the very significance of the, *'divine words'*, planted in the human mind as an immunological means of healing and repair, may be obscured and altered to justify deeds of bestial behaviour and inhumanity to man, such as we are witnessing in the Middle East, in direct contravention of Christ's advocation:(*'You must love your neighbour as yourself': & 'The lamp of the body is the eye: It follows that if your eye is sound, your whole body will be filled with light. But, if your eye is diseased, your whole body will be all darkness. If then, the light inside you is darkness, what darkness that will be!'*: *Matt. 5. 43-48. & 5. 22-24.*

Harmful concepts to be questioned: God punishes human sinful follies and misbehaviour by visiting the sins of the parents upon the children? If that were true, there would be nobody left. Plato, with his towering cognitive capacity to see into the heart of any conundrum, told us that over 2,000 years ago: *('The World is full of error')*. Jesus Christ our Saviour, known to all who came to hear and know Him, as *'The Healer'*, left us in no doubt about how the rights and wrongs of human behaviour would be judged: *('Yes, if you forgive others their failings, your heavenly Father will forgive you yours; but if you do not forgive others, your Father will not forgive your failings either'*:Mat.6.14-15): (' *Then Peter went up to Him and said, 'Lord, how often must I forgive my brother if he wrongs me? As often as often as seven times?' Jesus answered, 'not seven, I tell you, but seventy-seven times'*: *Mat. 18. 21-22).*

Therefore, we have been warned that *'whatever we do to our fellow men, we do to ourselves'*. In this respect, it must be understood that it is not our Creator who judges our behaviour. That task has been placed, as an obligation to *'know us'*, in every human mind, for the purpose of monitoring and regulating our behaviour and actions, for our own good and the well-being of all mankind. To think of the *'Supreme Creative Mind'*, reflected in the wondrous network of interacting dynamic systems by which Life and all living things appear to have been brought into existence, as uncaring, would be an absurd belittlement of *'The Word and Will of God': ('Does the pot say to the potter, 'Thou hast no skill!': 'Godless men will come trembling to the reckoning of their sins, and their crimes, confronting them, will accuse them. Then the virtuous man will stand up boldly to face those who have oppressed him, those who thought so little of his sufferings. And they, at the sight of him will shake with coward's fear, amazed he should be saved so unexpectedly. Stricken with remorse, each will say to the other, with a groan and in distress of spirit: "This is the man we used to laugh at once, a butt for our sarcasm, fools that we were!. The breath of Omnipotence will blow against them and winnow them like a hurricane. So lawlessness will bring the whole earth to ruin and evil-doing bring the thrones of the mighty down': Wisdom: 5. 1-32).*

(Gk. sarkasmos from sarkazein= tear flesh, gnash the teeth, speak bitterly).

Christ also made a point of telling us that He did not come into the world to condemn, but to save the world, reiterating that God has delegated the duty of giving an account of behaviour and actions to each human soul, so that each will receive reward or punishment according to their deserts: *cf.* The ancient Greek commandments, *'Know Yourself' & 'All things in Moderation'.* Does this not also affirm Mitchell Feigenbaum's hunch, *('Los Alamos Laboratory'),* that there must be a, *'A Universal Principle of Order',* inherent in the functioning of all world systems. This poses a problem of judgement for every human being, because we are all naturally *Geographers,* in search of the optimum location, to provide a good and prosperous life for our families in future.

The challenge we face, is to keep in mind that *'we share a world together as a family home'*, which inspires a higher cognitive faculty of appreciation towards ourselves and the world, bound together in a mystically beneficent state of existence. That view of Life and the world generates spiritual qualities of *'Love and Ecological Wisdom'*, in the human mind. Thus, the language of geographical knowledge can be read as a semiotic code, carrying true meaning from age to age, just as Galileo thought of the, *'Universe written in the language of Maths'*.

What appears to have been clearly understood in ancient wisdom, seems to have been forgotten in our modern contentious world. *(' The word of God is something alive and active: it cuts like any double-edged sword, but more finely: it can slip through the place where the soul is divided from the spirit, or joints from the marrow; it can judge the secret emotions and thoughts. No created thing can hide from Him; everything is uncovered and open to the one to whom we must give account of ourselves'):* St. Paul: Letter to the Hebrews: 4. 12-13.

Ponder these sagacious words, spoken by a highly intelligent mind, a linguistic polyglot, versed in all the idiomatic nuances of ancient Hebrew languages, with the ability to translate complex concepts into Greek, Latin, and other developing Indo-European languages- such as the Celtic tongue spoken by the Galatians, and it becomes evident that his understanding and view of the events unfolding around him had changed completely. What could have transformed his outlook, from that of an intense Zealot determined to destroy any of his brethren he considered to be a blasphemer, to that of one of the most ardent of Christ's Apostles? Certainly, with his knowledge of past calamities, recorded in the ancient Scriptures and folklore, as events of eschatological significance, emphasizing that the sins and misdeeds of man evoke the imminent wrath of God, he would have realized that Christ's gospel was very different. Disastrous events, such as the mass migration of a whole nation in the 3rd millennium BC, to form the ancient Hittite kingdom mentioned in the Old Testament; or the massive volcanic eruption of the Santorini caldera, could not be explained in this way- as expressions of the anger and wrath of God.

* **The vanished Hittite kingdom:** These people are thought to represent a mass migration from the North, either via the Balkans (peninsula bounded by the Aegean, Adriatic & Black Seas), and the Danube valley, in about 1900 BC. They eventually established an Empire centred on *Hattu,* situated on the barren Anatolian Plateau, dated about 600 to 700 BC. Why they chose this relatively unproductive granitic area to found a kingdom, remains a complete mystery. Equally mysterious, this highly cultured nation, speaking a sophisticated Indo-European language containing the roots of languages spoken today, and with the ability to engineer advanced irrigation systems, carrying water from distant from mountains far away, able to transform a barren region into fertile productive land, suddenly vanished from history. All knowledge of their location was lost, until 1834, when a French explorer, searching for the lost Celtic city of *Tavium,* came upon the ruins of the ancient city of *Hattusa,* containing the remains of a great Library, consisting of over 30,000 clay tablets inscribed with records written in cuneiform and the mysterious language code of *Linear B.* This was eventually deciphered and found to be a sophisticated Indo-European language, which has provided the roots of languages spoken across the Western and Middle East today.

Their decorated vases show that bull-leaping was a popular sport, which may have been part of a religious ritual symbolizing the perpetual struggle between good and evil in the human psyche, on the journey through life. The same theme is predominant in the ancient Scriptures, centred upon Man's fall from grace, in the *'**Mind of God**'* for disobeying the command given in *'The Garden of Eden', where the World and all living things were brought into existence, in a state of perfection':* **'Do not taste the fruit of the Tree of** Evil! For this *'Original Sin',* we are told that *'Adam and Eve',* and their progeny were expelled from their perfect world state of *'Eternal Life',* condemned to wander in a world fraught with hazards of their own making, until *'Humankind's deliverance from sin and damnation by the atonement of Christ'.*

* Their problems then compounded, when Cain killed his brother Abel, in a fit of jealousy over the inheritance of their father's estate. Thus, another divisive theme became prominent in the Old Testament writings, as *'Mortal sins of Envy and Murder',* closely associated with human quarrels and disputes connected with the acquisition of wealth and power. It colours the *'Abrahamic Saga',* the story of Abraham's exit from the ancient city of *'Sumerian Ur',* in about 2,000 BC, according to the accounts written in cuneiform and containing *'Amorite'* language, words of ancient Hebrew origin.

(Gk. Ioudaious from Amaraic:Yehudi from **Yisrael**= **'he that strives with God', the name that was conferred upon Jacob:** Ysraeli= *Israelites, The Jewish People, the 12 Tribes of Israel).*

* **The Santorini Caldera Catastrophic Explosion:** In about 1600 to 1700 BC an enormous volcanic explosion took place, when the vast caldera, (underground chamber), of **Santorini** erupted in one of the greatest volcanic events that has ever occurred. Estimated as over 7, on the Richter scale, it destroyed the ancient prosperous trading settlement of *Thera,* when the contents of the magma chamber increased by over 2,000 times in volume, due to seawater mixing with the red-hot magma, which caused a mighty eruption, estimated to have been about 29 cubic miles in volume. Known as *phreatic* eruptions, (water mingling with red-hot magma), such extremely destructive volcanic events are known to have occurred many times in geological history. As a result, it gave rise to a *Tsunami* wave over 30 metres high, which swept across the Mediterranean region and beyond, completely obliterating the prosperous ancient Greek Mycenaean civilization, described by *Homer* in his epic poems relating to *The Trojan War: ('Iliad' & 'Odyssey', written about 750 BC).* The story of these events is written also into the rocks, which tell us that a similar mighty explosion had occurred there 22,000 years ago. We know therefore that the Mediterranean region has not always been the comfortable idyllic environment, we have come to know during the last thousand years. We are reminded yet again, that nothing we think we know, perceive, or assume to be fixed, as *'Laws of Physics, Chemistry, Biology, Geography',* or anything, can be thought of as unchanging.

The world environment must also be regarded as constantly in a state of flux and change, at a greater scale of magnitude. Thus, the very rocks of the Earth, upon which we stand, together with the air we breathe and the seas upon which we sail and fish, must be considered as in a state of fluid change, not as permanent and material entities. The same reasoning holds true of living beings, as macro-molecular assemblies of properties, and especially of the mysterious cognitive faculties of the **Human Mind,** with the capacity to make sense of geographical experience, as ***'realities of existence related to the abstract dimensions of space and time'.***

Deleterious effects of the Santorini trauma: As a result of this traumatic volcanic eruption, the geographical environment of the Island of ancient Thera was drastically changed. The ancient trading settlement of *Akrotiri*, which was buried under 60 metres of volcanic ash and lava, has been excavated, so that it is now possible to walk along ancient streets, peep into houses, and inspect ancient amphorae containing the dried up remains of olive oil, fruit and grain, which were stored there in about 1600 BC. However, we cannot relive and regain ancient knowledge relating to social systems of organization, government, language of communication, sense of ethnic identities, trade routes, quality of literacy, or technique's underpinning the advanced *Bronze Age* civilization, which existed in *Mycenaean Greece.*

All were catastrophically disrupted by what must be one of the most dystopian natural events to have occurred on Planet Earth. We can only imagine what life must have been like at that particular moment in the '*spatial and temporal matrix of existence'.* The geophysical environment was also drastically changed. For example, the beach deposits now consist of black pebbles of a basaltic type of rock, and the sea temperature is markedly colder than is usual around other Greek islands, because of the greater depth of sea. The chemical cauldron below is still active. In the 1890's, it was reported that a fishing boat was found with the unfortunate crew boiled alive as they were fishing offshore.

We are reminded therefore, that environmental conditions in our planet Earth cannot be considered as fixed and unchanging, although other geological evidence tells us that the chip of an ancient continent, defined as the British Isles and *'The UK',* must have been located in an equatorial zone for episodes lasting over 50 million years. Otherwise, it would not have been possible for the formation of rocks, such as over 1 km of chalk during the Cretaceous Period, when gigantic creatures like the dinosaurs roamed the Earth, to have taken place. Such a formation, consisting of a steady rain of fine *coccospheres,* could only have taken place in a calm tropical warm sea, lasting a vast period, in an unchanging warm climate. This presents a profound mystery. We know from geographical experience that all we perceive, concerning *Life Processes and our World Environment,* confirms that all must be considered as interacting systems, in a fluid state of flux, and yet there arc anomalous events which defy explanation. Even the polarity of Planet Earth has changed several times in geological time. If we find it a challenge to understand what is so near, how can we place so much confidence, in what is happening at a stupendous magnitude of distance, far away in time and space? For example, (*'The coalescence of 2 Black Holes in the middle of a Galaxy, 300 X trillion billion miles away'*: BBC News:7 October 2020). Although attributed to outstanding scientific minds, acknowledged as *Polymaths,* commanding many branches of knowledge, such reports of discoveries beyond human understanding, concerning infinitely greater mysteries of creation, would be best left open to question. The ancient Greek philosophers warned their students to avoid the delusion of *'hubris',* overweening pride.

(Gk. Basanithes: lithos= stone: basanes= touchstone: Kokkos= grain, seed, tiny calcium carbonate platelet of marine flagellates.

Anomalia from homalos: anomalos= dissmilar, uncomfortable. Polymath from manthanein= to learn: polymathes= having learned a lot, with cognitive faculties of a high order: dystopian=ruinous condition, from dystopia= a very bad place).

The enigma confronting the Jewish scribes: For the Jewish scribes, given the task of recording the *Scriptures,* relating to the *'Covenant with God the Creator'* (ca. 160-70 BC), expressing their belief that they had been especially blessed by God as His chosen people, it must have been equally perplexing to make sense of calamitous natural events, such as the mass *diaspora* of a whole nation, (ca. 1900 BC), and the calamitous Santorini explosion, (ca. 1600 BC). The huge *Tsunami* wave, which swept across the Mediterranean Sea, following those calamitous events would have obliterated the ancient records held in the old Hebrew Library, located in the original Temple built by King Solomon, dating back to 850

BC, leaving only garbled memories of *The Great Flood of 2,000 BC.* Also, *The Santorini trauma* would undoubtedly have compounded the problem they faced, in their attempt to reconcile destructive natural events with the concept of a loving and caring God: especially since the flourishing *Mycenaean Civilization,* which had stemmed from, '*Judaeo-Phoenician roots*', together with all the systems of human organization on which that highly prosperous society was based, had been entirely obliterated. Perhaps that would explain why they evidently chose the garbled memory of the older *Noah/Gilgamesh Flood,* to justify their claim to be the '*Chosen elect of God*': *(Aromaic Yehuday: Hebrew Yehudi from '*Yisrael= **he that strives with God***': Book of Genesis: 32. 29.

The impact of the destructive seismic sea-wave, towering up to 30 metres in height, would have wiped out all settlements, leaving very few people capable of recording the event, and most probably with the effect of reducing the former prosperous *Mycenaean and Minoan Bronze Age,* way of life, to a '*Neolithic*' (Stone-Age) level. In fact, the extraordinary prosperity and sophisticated nature of that highly developed civilization was not known, until Sir Arthur Evans discovered the Palace of Knossos in Crete (1900), and the ruins of the ancient city of Troy *(Ilium),* south of the Dardanelles, were excavated by the Archaeologist Heinrich Schliemann in 1894: *('Odyssey, & Iliad': Epic Poems by Homer: written about 850 BC).* Schliemann also discovered the heavy gold necklace attributed to *Helen of Troy,* which has since gone missing from the Museum in Berlin, where it was placed on display. The pictures uncovered on the walls of the excavated Palace tell us much about the advanced cultural level of Mycenaean Civilization.

For example, judging by the elegance of the ladies depicted there, it is very evident that women were highly respected, foremost in society relationships, and certainly not downtrodden, as we see today in so many society relationships of our modern world. This was most probably a highly cultured '*Matronal*', democratic system of organization and government. They had developed highly efficient irrigation and sewage systems, with water flushed toilets, superior to the sewage disposal systems, encountered in holiday accommodation facilities available today, throughout the Greek Islands. Could this be the land of '*Euphrates*', the fertile valley where people must have lived in peace and harmony and mutual brotherhood, for centuries in the distant past? Place names tend to reflect geographical experience, but the land along the river we now call '*The Euphrates*', is riven by hateful conflicts, turbulence, and abject disorder, just as it was described by the Jewish scribes in the old Judaic Scriptures.

Therefore, the Genesis account of the '*Great Flood and Noah's Ark*', represented by the Jewish scribes as a sign of God's disfavour, must be regarded as a fictitious or allegorical attempt to explain how the '***Will and Word of God***' works to regulate human behaviour and systems of human organization. However, geological studies have confirmed that the '*Great Flood*' *which* took place in 2,000 BC, as recorded in the '*Gilgamesh Epic*', was most probably caused by a catastrophic stormy episode of '*climate change*', at an interglacial phase of climatic warming within '***THE ICE AGE***', commencing about 10,000 years ago, with rapid melting of glaciers and ice-caps, and the formation of a vast fresh-water lake. As a result, an enormous flow of water broke through a low range of hills, south of the Dardanelles, flooding what had been a highly prosperous and densely populated basin, which is now the Black Sea. Studies of the seabed have revealed the ruins of drowned towns, as well as evidence of highly advanced agricultural organization, characteristic of the remarkably sophisticated level of civilization achieved in mainland Greece, during the '*Early &*

Middle Helladic Period' of the Middle Bronze Age: (ca. 1800- 1600 BC). However, it is very easy to misinterpret bits and pieces of evidence relating to past geographical events and systems of human organization, and to make false assumptions, which then are transmitted as dubious information, readily accepted as valid knowledge.

More recent archaeological discoveries, associated with research into the sequent occupancy and systems of human organization, connected with the *'Minoan'* settlement around Knossos, have revealed that the prosperous way of life achieved during the *Mycenaean Era,* may not have been entirely obliterated by the *Santorini cataclysm.* On the contrary, it appears that *Mycenaean Knossos,* may still have been prosperous and thriving as a centre of a trading network extending as far as India and the Far East, at the time of its destruction, now thought to be in 1,370 BC. It seems likely that the cause of its decline may well have been the result of a bitter conflict with a rival *Mycenaean* power, located in the Argolid Peninsula of ancient Greece.

(Gk: diaspora: diasperein from sperein=sow, scatter, disperse, people migratiing. Allegoria from allos= other+agoria: Allegoria =speaking, description of an event under the guise of another event or idea).

These events would have been held close in memory by the educated Scribes charged with writing the Judaic Scriptures in 850 BC. Why therefore, was a vaguely remembered story about a catastrophic flood, alleged to have overwhelmed the world over 1,000 years before, used to propagate the myth that such natural events may be explained as punishments imposed by God, for the wrongdoings of man. Seemingly, the more recent disastrous war, which had destroyed the prosperous and highly cultured *Minoan Civilization,* achieved by the descendants of those who had settled in the *'Land of Good Brotherhood (Euphrates),* had been erased from

204

memory. The ancient Hebrew records make no mention of it, or of the destructive consequences that must have wrecked the harmonious relationships, prevailing in the *Minoan* political and trading systems of organization. Perhaps the memory of that hateful conflict, motivated by greed and acquisition of power and wealth, sat uneasily in the conscience of the pharisaic ruling junta at that time. After all, they would have been aware that they had profited as usurpers of the Minoan wealthy trading empire, established by their '*blood brothers*', living in peace and prosperity, and enjoying a superior cultured way of life.

How could that transgression be reconciled with the *Mosaic Law,* handed down to Moses in 1450 BC, which commands that you must love your neighbour as yourself, reflecting the love of God for man and all creation, as well as regulating the behaviour of the priesthood? Desire for worldly things, invariably incites hostile divisive relationships, culminating in cultural decline. That is how naïve concepts may be seeded in the human mind, to become a misleading interpretation of the way that '*the Word and Will of God*', works in the world. For example, whereas archaeological and anthropological evidence shows that women were rightly honoured with high status in *Minoan civilization, 'The Judaic Scriptures',* as recorded in the Old Testament appear to have relegated womanhood to an inferior and subservient status in society. Christ left no room for doubt in this context, when he was questioned about the right to divorce a wife, or to have several wives as was permitted under Jewish Law, as if women were mere chattels: *('Everyone who divorces his wife and marries another is guilty of adultery, and the man who marries a woman divorced by her husband commits adultery')*: *Luke 16. 18.*

Many other erroneous and misleading concepts of this nature have been propagated in this way and continue to erode social harmony and human relationships in the modern world. For example, *'The Book of Daniel in "The Prophets"'*, has proven to be a work of fiction, written 400 years after the events, containing spurious information, such as the survival of *'Shadrack, Meshac and Abednego'* after being thrown into a red hot furnace, by order of *'King Nebuchadnezzar'*, seeding the idea that **God our Creator** intervenes to regulate human behaviour by works of magic: *Daniel: 3. 1-40.*

False alliterative, allegorical, or metaphorical assertions of this kind, tend to proliferate, and to inflame more destructive divisions at a world level. Thus, rebellious factions, such as *'Women's Rights'*, and *'Slavery Riots'*, are being inflamed by rallying slogans, such as, *'Break the glass ceiling!'*, and *'Black lives matter!'*. Protest movements of this kind, contain mixed messages attractive to political and other opportunists on the lookout for gain, which may have the effects of nullifying righteous causes, *(e.g., 'Women's Rights')*, as well as inflaming past vengeful recriminations and causing more destructive disorder in systems of human organization. Paradoxically, just and unjust causes can be conflated in that way, to become a highly disruptive dystopian force, capable of cascading into extreme disorder in world systems of government and organization. Indeed, do we not see this scenario unrolling in the Middle East, the place once hallowed as the *'Holy Land'*, where once people lived in harmony and mutual brotherly love, but is now torn by acrimonious hate?

The folly of Man and the delusion of a material world full of riches: Three hundred years ago, a second in geological time, the world would have looked a dismal place, where most people were condemned to work as slave-labour, riddled with disease and struggling to survive, on the brink of starvation. Eventually the inequality between the poorest and richest became so unbearable, that the mass was goaded to rise in vicious rebellion, culminating in barbarous hostilities, such as happened during *The French Revolution,* and *The*

American War of Independence. At that time, no one could have imagined that the vast sandy deserts of the Middle East contained a source of immense wealth, in the vast underlying oil deposits which were discovered in Persia (1908), and later in the sandy wasteland deserts of Saudi Arabia, in 1938. These oil deposits represent the fossil remains of vast numbers of animals and plants that have lived during the last 500 million years, including the gigantic dinosaurs which came into existence during the Mesozoic Era, and roamed the Earth, when it became covered with luxuriant forests and vegetation at the time of the Cretaceous (chalk formation), and Jurassic Geological Periods In 1817, when Shelley, with his wife went to Thebes on a grand tour, to study the archaeological discoveries of ancient Egyptian monuments, he could not have guessed the significance his poetic musing would later acquire, when he saw the remains of a huge statue buried in the sand:-

'My name is Ozymandias, King of Kings. Look on my words ye mighty and despair, nothing beside remains. Round the decay, of that colossal wreck boundless and bare, The lone and level sands stretch far away.': ('On seeing the half-buried remnants of a huge statue of Pharoah Rameses II, 3ʳᵈ Century BC: Percy Bysshe Shelley).

No doubt he was thinking of the slave labour, as well as the waste of all the effort that must have been needed to make such a colossal sculpture, and all for nothing. He could not have guessed that just beneath his feet lay a potential source of immense wealth, which a century later would be fought over, in a savage desert war, with hideous loss of life. The same people who had mastered the skills and organization to construct the greatest architectural works in the world, such as the Pyramid at Gizeh, were also prepared to destroy their greatest resource like beasts of prey-- their fellow human beings. We may begin to see the significance of the ancient Greek philosophical strictures: **'Know Thyself'**, **and** *'All*

things in moderation', refined by Christ as, **'You must love the Lord your God with all your heart, with all your soul, with all your strength, and with all your mind, and your neighbour as yourself':** *Luke 10. 37-28).*

Thus, we may learn that a materialistic perspective of the world leads to an exploitive outlook, conducive to regimes of *'tyrannical slavery'*. Does this not epitomise our modern world, where reckless exploitation is destroying the environment, and *'globalization'* has come to mean *'trawling the world for cheap labour'*, regardless of human life and well-being. Christ's words warned us of the consequences, when speaking about the judgement we may bring upon ourselves, of following a way of life devoid of the spiritual qualities of **'love and care'. ('I tell you solemnly, in so far as you did this to one of the least of 'my adelphoi'** *(brothers)* **of mine, you did it to me; ----in so far as you neglected to do this to the least of these, you neglected to do this to me':** *Matt. 25. 31-46).*

(Gk.muthos= myth, traditional story embodying some event or religious belief:

Pharisee from Gk. Pharisaeus, from Aramaic pnsayaa & Hebrew parus= separate, superior group:

Enigma from Gk. Aenigma, from ainos=fable from ainissesthai= to speak obscurely:

Paradox from para=against + doxa =opinion+ contrary to belief, incredible, absurd:

Dinosaur from Gk. dino= fearsome+sauro= lizard, terrible monster:

Gk. Ozymandias=a tyrant, dictator, megalomaniac).

Convergence of Science and Theology: Significance of geological evidence

We know, from the records written in the rocks and in the geomorphological features of our Planet Earth, that calamitous natural events have occurred frequently, in the course of geological time. For example, the Santorini explosion pales into insignificance compared with the enormity of the explosion of the *'Yellowstone Super volcanic Caldera'*, 640,000 years ago, when the entire contents of the magma chamber covering an area of 34 x 45 miles in extent, amounting to hundreds of cubic miles of incandescent lava, ash and pyroclastic gases were ejected into the atmosphere. All life existing at that time would have been obliterated over a vast area. This will inevitably happen again. Perhaps next week, or in another million years. Already the Earth surface in Yellowstone National Park has risen to over 8,000 feet, in a region ringed by mountains over 10,000 ft. in altitude, due to the build-up of pressure deep in the mantle of the Earth's crust. In the event of another eruption on this catastrophic scale, all the geophysical systems of our home-world planet would be affected, including the composition, structure and circulatory systems of the atmosphere, oceans and in the mantle of the earth. All their interactions would be drastically altered. Already there are signs of increasing turbulence within the mantle, with more frequent explosive eruptions of geysers and toxic gases. The most alarming recent discovery concerns the emission of the very rare gas, *Helium- 4,* which can only be produced by the decay of *Uranium and Thorium,* *('US Geological Survey:2020).*

Therefore, we must consider that the *'crisis of climate warming'*, now facing world, may indicate that the Earth is warming up from within, possibly as a consequence of nuclear fission reactions. Recent research has revealed that the Earth, viewed as a dynamic fluid system, may be entering a phase of extreme turbulence, in which the rocks of the mantle, down to the *'Moho unconformity'*, appear to be warming rapidly because of exothermic reactions below. *('There are signs that the crust of the Lithosphere in Greenland appears to be thinner by 80-100 km., than the average thickness of 280 km., recorded over the Earth'*:

'Erosion by Earth's Central Heating': NS: 17 Aug. 2013; V 2930 p.16: Alex Petrunin). **(75).**

If this should prove to indicate a depletion of the crust, by chemical transitions in the asthenosphere, it would confirm that a scaling up of turbulence has occurred in Earth's dynamic Geomorphological system. We already know that the geomorphological features of our planet Earth, as we perceive them, are very different from those recorded in the petrological evidence contained in ancient rocks. For example, the composition of rocks in the British Isles, ranging from igneous rocks of ancient *Caledonian Orogen,* (mountain formation), in the Scottish Highlands, and the very ancient *metamorphic rocks,* (change by heat and pressure), of The Lizard at Lands End in Cornwall, to the gently folded stratigraphy of the Southeast and the Thames Basin, provide evidence of *'Continental Drift',* spanning over 500 million years.

We can even measure the rate of drift of parts of the Earth, such as the *'Atlantic and Arabian Plates',* in terms of millimetres per year, and we know that India was once joined to Australia, but for some reason, has move northwards at astounding speed, judged by the scale of *'Time',* in terms of geological chronology. The behaviour of seismic shock waves gives reason to believe that the core of the Earth may be solidifying into gigantic iron crystals, which could mean that phase-transition is taking place, as the swirling nickel-iron core changes to a solid state, because of a spike of energy resonance in Earth's inner chemical cauldron. The rapid rate of warming during the last 50 years, leading to the *'Crisis of Climate Change'*, could also be due to increasing chemical reactions within the interior of the Earth.

Although the environmental conditions linking the oceans, atmosphere, and structural geomorphology of our planet, have changed dramatically, with episodes of violent turbulence lasting millions of years, unsuitable to support the existence of *Life,* our planet has never been cooler or a more pleasant *'home-world',* than we now perceive it to be. For billions of years, it has been an extremely hostile geophysical environment, when it has been far too cold *('Ice-ball Earth);* or too perilous *('Toxic Atmosphere'* or *'Meteoric bombardment');* or too dry and parched with vast areas of desert totally devoid of water, the essential *'nectar of Life'.* Meanwhile the radiant heat emitted by the sun has been getting stronger. Yet, somehow Earth's geophysical environment has managed to remain potentially conducive, to a support for *'Life Processes'.* **The presence of Life in our Earth world-home is truly an astonishing mysterious phenomenon to behold:** *('It's time to take our eyes off the heavens, to look at something much stranger, Earth itself! Don't look up, Look down! To understand Life other planets, you need to look down at the rocks, as well as up at the stars. To learn about alien life may seem as sensible as reading tea leaves, but the history of our planet is written into its crust, and that history tells us much about the conditions necessary for Life'):* 'Lucky Planet': Ch.4: 'Constant Change': pp.47-63: David Waltham: Geophysicist & Astrobiologist. **(76).**

The most remarkable thing about our planet is that it has not become too hot like *Venus (420 C),* or *Mars,* where it is far too cold and stormy, with winds up to 400 mph., for life to exist. Therein, lies empirical scientific proof, which was confirmed when, *'The Nuclear Instrument of Death,* was in process of preparation at Los Alamos Laboratory, in 1945: **In** all systems there is an inherent **"Principle of Universal Order", which we now know how to employ, for good or evil, determined by the store of goodness or wickedness, held in memory by the human mind and heart.** In this light, it becomes clearly evident that all events observed as geographical experience, can only be understood as emerging probabilities, shaped and influenced by what has gone before, and is even reflected in the *Language* we speak, concerning recollection of the past. How reliable are out folk memories of past events, such as *'Great Floods'*, and *'Pandemic Plagues'?* Questions concerning the mysterious *'Life Force of Biogenesis'*, now command the urgent attention of us all, especially Theologians and those Scientists who have begun to question the assumption that, *'mathematical relations of the so-called laws of nature are the most basic level of description, but rather if information is regarded as the foundation on which physical reality is constructed':* *'Information and the Nature of Reality':* *'Introduction: Does Information Matter?':Paul Davies: ('Life is so extraordinary in its properties that it qualifies for the description of an alternative form of matter. Most textbooks focus on the chemistry of Life-which chemicals do what inside the cell. Obviously, Life is a chemical phenomenon, but its distinctiveness lies not in the chemistry as such. The secret of Life comes instead from its informational properties—A living organism is a complex information-processing-system'):*Paul Davies: *'The Fifth Miracle: Search for Origin and Meaning of Life').***(77).**

Theologians have been conveying this optimistic message of hope from time immemorial, as the *'Word and Will of God, in dialogue with the Human Mind'*. Scientists, formerly inclined to deny the existence of God, and to favour explanation of *'the inception of Life in a beneficently ambient world'*, as a matter of random chance, have been left little option by the discoveries of *'The Genetic Code'*, and

'*Quantum Electrodynamics*', but to acknowledge, albeit somewhat half-heartedly, that we are in communication with a '*Supernatural Creative Mind*'. The onset of turbulent disorder in world systems, characterized by calamitous events, such as '*Climate change*' and '*Pandemic disease*', have made us take notice, '*that complex systems such as living organisms, societies and human beings can no longer be explained by assuming a reductionist view of ourselves and world events, in terms of material components and their chemical interactions.*

Confusion over explanation of natural events as the Will of God: With light of hindsight, it would appear that the learned Jewish scribes engaged in writing and recording the Scriptures concerning the '*Covenant between Israel and the Word of God*', in about the 6thCentury BC, may have been equally perplexed by the extraordinary events that had taken place, with disastrous effects upon their lives, such as the traumatic impact of '*The Great Flood*', and the consequent mass migration of whole nations taking flight in terror. How could such terrible sufferings be reconciled with their concept of a '*loving God?* They would have been thinking in terms of linear relationships, by attributing the state of the world directly to the judgement of God, whereby immoral and wicked behaviour would invite the wrath of God, with the onset of traumatic events. However, with the benefit of advances in scientific knowledge, this naïve mindset is no longer tenable. Christ made it very clear that **the Mind of God does not work in this way:** *('For God sent his son, not to condemn the world, but to save the world—to redeem the sins of the world'): John 3-17.* Also, since '*Living systems are cognitive systems*', and **living,** viewed as a process, is also **a process of cognition,** we are able to know that a living organism tends not to react to environmental influences '*through a linear chain of cause and effect*', *but to respond with changes* (adjustments), '*in its* <u>*non-linear closed*</u>

autopoietic network, enabling the human mind with the power to bring forth an inner world, intimately linked to language, thought and consciousness'): 'Santiago Theory of Cognition': 'Biology of Cognition': Humberto Maturana :1978. **(78).**

Chaotic effects of increasing turbulence in Geophysical Systems: Our precious *Biosphere* consists of a very thin gaseous layer (2.5 km), of breathable air, a fragile tenuous mixture containing Oxygen in just about the ideal ratio, as an ideal support for Life: *(%* by V=21: % by W= 23: Parts per M=209,445: Molecular W=32).

We know only too well from geographical experience, that it can become polluted and intoxicated by reckless human activities, involving exhaust emissions of smoke, nitrous oxide, methane, CFC refrigerant gases, etc. In addition, the knowledge gained from study of *'The Photoelectric Absorption Effect',* has shown that fluid systems, when exponentially energized, eventually break into patterns of mounting turbulence, which can cascade into unpredictable patterns of extremely destructive disorder. In this connection, has any thought been given to the probable *'barrelling effects',* of fast-moving vehicles and rocket projectiles driven at the escape speed of 37,000 mph. by waste gases, along the circulations and through the structural shells of our **Earth world?**

(Biosphere, Atmosphere, Troposphere, Ozonosphere, Stratosphere, Magnetosphere): Recent research has revealed that these vital gaseous protective structural shells are far more dynamically active, than atmospheric scientists had previously realized: e.g.:-*'Baldwin, M.& Dunkerton, T.: 'Journal of Geophysical Research:' Biennial, Quasi-biennial & Decadal Oscillations of Vorticity in the Stratosphere:DOI:129/97:jD62150.* **(79).**

Earth's Protective Shields: We know it to be our good fortune that our planet is equipped with a protective barrier, in the lower part of the *Stratosphere (15-35 Km),* consisting of a relatively higher concentration of Ozone gas (O^3: at 10 pp million), compared with the average atmospheric concentration of 0.3 pp m, which acts as a protective shield to the potentially harmful effects of the Sun's UV radiation. Without this protection, it would be improbable for Life as we know it to exist, because of the genetic damage caused by exposure to the searing *'Photoelectric' effects'* of sun's UV light. Does anyone know what might happen, should a vast hole be blasted through this vital, but flimsy and nebulous gaseous shield, by the mighty onslaught of rocket after rocket, travelling at 37,000 mph.? As for the *Magnetosphere,* so little is known about its origins, let alone how it interacts with the networks of the *'Web of Life Processes',* as well as its mysterious connections with *Geophysical systems,* and Geomorphological formations. We do know, for example that migrating birds appear to make use of magnetic fields, to navigate their way, travelling a long distance on the way to winter nesting grounds: *'Quantum Biology Comes of Age': J. Khalili'.***(80).**

The Scalar Enigma of Geographical knowledge and experience: Fig.1: The timeline depicting the *'Geophysical Process of Oxygenation of the Atmosphere',* must be understood as a sketchy representation of inferred events, which culminated in the structural formation of the Earth world, as a cradle suitable for the emergence and support of Life. Note that it purports to cover 4·5 billion years (assumed to be the span of Earth's existence), but little can be learned from the petrological evidence in the rocks, relating to events beyond about 500 million years, *(The Cambrian Period, the earliest of the Palaeozoic Era),* when the first signs of living organisms may be detected. We can be sure that the

Geomorphological Structure of our planet has constantly changed: *Magnetic Polarity* has flipped many times.

The British Isles represent a chip of an ancient continent *(Pangaea),* once in the southern hemisphere, which has travelled across the climatic zones to the location we know, as the *Northern Mid-latitude;* and the *Mediterranean Sea* has dried up and flooded again many times. We can only use these bits and pieces of unreliable information, to guess all the myriad interactions which may have contributed to the astonishingly ordered ambience of our Earth-world, to make it such a benign environment. We can only reconstruct such a complex network of connections, in terms of a linear pattern of events with time. However, in the light of *Q Electrodynamic revelations,* we have learned that the ordered patterns perceived as emerging events, can only be explained as a far more mysterious and ingenious chain of *'teleonomic communication',* __involving periodic linear reactions, interspersed & linked by contingent aperiodic interactions of knowledgeable adjustment.__ *('The fundamental feature of Quantum Electrodynamics is that energy and frequency are intimately related—this Constant of Proportionality, 'alpha', is central to our entire description of the Universe, and determines how much energy there is in empty space'):* Michael Brooks: *'13 Things That Don't Make Sense'.* **(81).**

It would appear that the Human Mind is receiving a message of fundamental consequence, riveting attention upon the deteriorating state of the world.

Climate Warming Crisis: Consider the consequential probabilities of this train of events:-

* The wasteful demands of 2 World Wars led to an unbridled development of mass industrial production, requiring vast amounts of coal, iron and steel, to supply vehicles, tanks, battle ships, planes, weapons and armaments. The remains, in the form of toxic debris, detritus, rusting bombs and poison

gas containers, now litter the oceans and seas across the world. Production of armaments has continued apace, with the storage of weapons of immense nuclear destructive power in underground bunkers and silos full of guided missiles.

* Pollution of the atmosphere, oceans, seas, rivers, and land surfaces has continued unabated, with poisonous material of all kinds, including nuclear waste that no one knows how to dispose of safely.

* * The development of nuclear weapons has involved the testing of explosive devices in deep bores across the world, in remote regions, such as the Nevada Desert, Kazakhstan, Siberian and Pacific Islands, resulting in continuous spikes of unimaginable energy, into the *'homeostatic interacting Geophysical Systems' of our wonderfully ordered Earth-world.*

* * * * There are signs that seismic activity is increasing, notably along *'The Pacific Ring of Fire',* particularly in New Zealand, Chile and Yosemite, and volcanic vents which have remained dormant for a long period are also showing signs of increasing activity: e.g. Iceland, *'Where a volcano, dormant for 800 years, is becoming active: it could herald a period of eruption lasting centuries':* Magnus A. Sigurggeirson: ISOR: Iceland Geosurvey. **(82).** * * The Mantle in Greenland appears to be thinning, by about 80 km., which could mean that the Earth's interior is warming up, probably from energy released by nuclear fission. In 2010, when **Eyjafjallajokull, Iceland,** became active after a long period of dormancy, thick black clouds of fine ash spread over Europe, at high level. Airlines were forced to cancel 100,000 flights, because of the hazard of jet-engine failure, at an estimated cost of €2·5 billion. We are reminded that the Earth (as the home of man and all living things must be viewed and valued as interlocking and interacting fluid systems, reflecting the *atomic/molecular/shell/spin/structure,* of molecular

Systems. Viewed in this way, Earth's shell-like geomorphological structure is not fixed and immutable, but revealed, as a chemically active cauldron, in an astonishingly quiescent state of equilibrium. *('Homeostasis':Core/mantle/asthenosphere/lithosphere/biosp here/atmosphere/troposphere/ozonosphere/stratosphere/magn etosphere),*

With knowledge of such critical importance, who in their right mind, would risk conducting dangerous experiments, such as testing explosive nuclear bombs in deep bores, like the Tsar Hydrogen bomb (1962), of such immense hideous destructive power that it shocked the scientists involved. The consequential effects upon the Earth's regulatory internal systems, by energy released and injected into the chemically active core *(at 8,000 ° C),* can only be imagined, and are yet to be experienced. With the knowledge of what happens in highly destructive *'phreatic'* volcanic eruptions, such as the *'Santorini Caldera'* eruption, when pressure in the *'magma chamber'* increased catastrophically, (X 2000), caused by a *'phase state change'* of water mingling with molten magma, who could be so foolish as to test nuclear explosive bombs in close proximity to *'Earth's chemically active cauldron',* at the same temperature as the surface of the Sun?

('The number of nuclear weapons in the world has declined since the Cold War, from a peak of about 70,300 in 1986 to 14,550, according to the FAS (Federation of American Scientists), but the pace of reduction has slowed – the perceived value of acquiring nuclear weapons has gone up, while the repercussions of violating treaties has declined— the new arms race has already begun. It's different in nature from the Cold War, which focused on quantity, and 2 super-powers, producing absurd numbers of weapons. Today it's focused on quality and involves several nations— the risk for nuclear conflict is higher than it was during the Cold War: W.J. Hennigan: 'Time'. (83).

Nuclear Nations: Warhead Stockpiles (2018): UK: 215: France: 200: USA: 6,800: Russia:7,000: Pakistan: 140: India: 130: Israel: 80: China: 270: N. Korea: 15.

Charles Darwin was ahead of his time, in comprehending the World's Geophysical Systems in terms of *'Turbulent Fluid Dynamics'*, recorded in his studies of *the Galapgos Islands & Coral Reefs,* noting the dynamic connections linking his observations, *(Biology+Geology+Plant & Animal Adaptations).* **(84).** He came to see that they could be explained as a result of uplift, subsidence and movement of vast areas of the Earth's crust on the ocean floor. Thus, he may be regarded as a pioneer of *'Plate Tectonics'*, a century before Alfred Wegener (1880-1930), who established the *'Theory of Continental Drift'*, explaining the fascinating geomorphological and biogeographical relationships between present continents, as due to the splitting up of a former super-continent, *'Pangea'*, which once stood in the southern hemisphere, about the latitude of Australia. Therefore, we know that our Earth-world can no longer be understood as, in a fixed material state of existence: but, rather in a constant flux of change, emerging and shaped by what it was before. For example, we know that the *'Indian Plate'* has also moved from a position in the southern hemisphere, at an extremely rapid rate, to collide with the *'Eurasian Plate'*, resulting in the uplift of *'Himalayan Mountain Range'*, and creating a barrier in the former patterns of circulation in the interactions between *'Oceanic+Atmospheric+Biospheric Systems'*.

Consequently, climatic patterns changed completely, with the knock-on effect of influencing changes in sequential patterns of *'Geographical Existence', for all living things.* Such phenomenally rapid movements may be attributable to the release of energy by *'run-away nuclear fission'*, in pockets of the Earth's mantle, causing powerful convection currents to form in molten layers, capable of driving a large crustal plate across the globe. These powerful movements of flow are thought to have been caused by the formation of massive *'mantle plumes of expanding molten fluid magma'*, which impact on the bottom of the Earth's Crust, with immense and prolonged extrusive force. The large area of igneous rock forming the region, known as *'The Siberian Traps'* for example, covering an area of over 3 million sq. miles, consisting of about 1 million cubic miles in volume of basaltic rock, represents one of the most catastrophic volcanic events, which have occurred during the last 500 million years. The rocks bear evidence that the extrusion of molten rock must have continued for over 2 million years.

Similarly, *'The Great Extinction'*, which occurred during the *'Permian-Triassic Period'* about 250 million years ago, causing the mass extinction of over 95% of all living creatures, is regarded as among the most destructive events in geological history. We now know that although the constituents of the oceanic floor and the continental crust *('the lithosphere')*, are very similar-chemically defined, the crystalline structure of rocks is subject to being re-ordered by changing parameters of heat and pressure. Thus, the oceanic floor consists of coarse-grained dark igneous rock, *'gabbro'*-similar to basalt, which forms the underlying mantle and is fine-grained, whereas the upper continental crust consists of lower density acidic rocks called *'sial'*-composed mainly of silica & alumina. The difference in density between oceanic floor (gabbro), and the continental rocks (sial), can be attributed to the proportion of *'silicates'* each contains: sial consists of 60%, whereas gabbro consists of 50%. Thus, the processes by which the continents appear to have been formed, may be understood as an intriguing pattern of *'contingent connectivity'*, **reminiscent of the contingent chains of knowledge-based protein messengers, discovered to be acting as regulatory systems, required to provide the precise specific parameters** (Temperature, Pressure, Density, etc.), **in every step of the way in living metabolic systems.**

The processes by which the continents may have been formed are discussed in the paper: *Jeff Hecht 'Rise of the Upper Crust': N.S. 8 April 2015: 3017: p37.* **(85).**

The extraordinary explosive tectonic movements, which caused the *'Indian Plate'* to break away from the *'Gondwana Supercontinent'*, comprising Africa+ Australia +Antarctica & South America, about 140 million years ago, involved even mightier *'magmatic extrusions'*, as the parts split asunder and moved away from each other at different velocities. The Indian Plate broke away from where it was joined to Madagascar and shot northwards to collide with great force with the *'Eurasian Plate'*, marking the unfolding event of *'The Himalayan Orogenesis'*, which is still continuing to this day. We know therefore, that our Earth world Home must be considered as quivering and shaking violently, and extremely vulnerable to any external activity, which might trigger the onset of even greater **chaotic reactions,** in what is already a highly turbulent system. Thus, the astonishing conclusion must be that the violent movements which caused the break-up of *'Pangea Supercontinent'* over 500 million years ago during *'The Precambrian Period'*, have continued ever since, as a process of *'dynamic geomorphological change'*, and is still unfolding before our eyes as an active *'orogenic event'*. Mt. Everest is still rising, as the *'Indian Plate'* plunges beneath the *'Eurasian Plate'*, creating a subduction zone, where *'exothermic kinetic reactions'* result in recrystallization and *'metamorphosis'* of the molten rock, leading to a phase-change of density and uplift of the Himalayan Mountain range.

It can no longer be taken for granted that our Earth world is robust enough to withstand any shocks that may be inflicted upon it, through aggressive human folly and exploitive uncaring greed, such as testing nuclear weapons and stockpiling warheads of vast destructive power. What good would it do our home-world, should underground silos packed with nuclear warheads suddenly explode, detonated by exothermal spikes emanating from Earth's chemically active interior? Although our knowledge of geomorphological change is limited to the last 500 million years, (10% of Earth's planetary existence), it is enough to show that for 90% of the time, environmental conditions have not favoured the emergence and existence of Life.

On the contrary, conditions have been extremely hostile for Life to come into existence. In addition, the Earth's surface is constantly in a fluid state of flexing, bending, and cracking, caused by *'Isostatic adjustments'*, thtough changes in weighting, connected with episodes of glaciation and orogenesis. For example, in Norway and Scotland the land is rising, following the melting of the heavy icecap during the inter-glacial period in which we are still living, whilst Southern England and the Netherlands are sinking, pivoting like a seesaw.

The atmosphere only came into existence, suitably oxygenated to support Life processes, 350 m y a, during the Carboniferous Period, and certainly cannot be regarded as in a fixed state, robust enough to absorb the repeated shocks of space rockets, thundering through its tenuous protective layers with a gargantuan destructive power of vorticity. Has any thought been given to the gigantic holes torn through the vital protective shield of the *'Ozonosphere'*, by such destructive onslaughts? Without the protection of this nebulous gas layer, consisting of less than *10 p p million of O^3*, between 20 to 25 km., which shields us from the lethal *effects of UV light,* Life processes would be untenable. By comparison, the horrifying loss of life world-wide, caused by the effects of *'Covid 19 Pandemic mutants'*, would probably rise to an horrendous rate of several millions per year, caused by the destructively searing effects of UV light upon living tissue.

('The most important of all systems for life—the atmosphere—can be unstable over an interval 100 million years long. Planetary atmospheres can change enough to cause mass extinctions and maintaining an atmosphere conducive to animal life for the staggering periods of time necessary for life to evolve and diversify may be the most difficult feat of all'): Peter Ward & Donald Brownlee: *'Rare Earth: Why Complex Life is Uncommon in The Universe').* **(86).**

How could it be, that Life should come into existence, in such an unpromising and hazardous geophysical system, as we now know our Earth-world to be? Also, informed by scientific discoveries, *(Organic Chemistry, Medical Research, Biogeographical, Psychology & Educational),* there can no longer be any shadow of doubt, that the intricately ordered interactions perceived as **'The Web of Life',** are astonishingly well organized and sustained, whereas the geophysical environment appears to be subject to episodes of catastrophic disorder, at a vastly different dimension of scale. According to the geological records, there have been 3 major disruptions in the connections linking Geophysical and Life systems,' *The Great Extinction' (252 mya:Permian-Triassic: 90%): 'The Manicouagan Impact '(214 mya: little effect on Life): The Great Dying': 65mya: 'The Chixulub Crater' Impact: 75%),* as well as major Glaciation events *('Ice-ball Earth').* Recent studies of *metamorphic* alteration in the mineral composition of the rocks, after the *Chixulub Impact* in the Mexican Gulf, confirm that planet Earth would not seem to be an ideal setting for the emergence of Life. The whole atmosphere would have been intoxicated by this cataclysmic event: *'Palaeogeography, Palaeoclimatology, Palaeoecology: Eric Tohver (Un of Perth): 'The Great Dying'.* **(87).**

In this event, the high probability is that the asteroid impact set off a series of earthquakes measuring 9·3 to 10·5 on the Richter scale, which shattered the underlying shales over more than 1,000 km., releasing a lethal emission of methane, estimated to be over 1,600 gigatonnes in volume. Methane is 23 x more polluting than CO^2 as a *'green-house'* gas, which raises the question whether world governments should be more concerned about the potential release of vast stores of methane known to be in the oceans, than with the *"carbon footprint"*, in making decisions to mitigate hazards associated with climate warming. A further study of the gigantic prolonged emissions of *'flood basalts'* in Siberia revealed that explosive vents had been driven through the rocks by volcanic emissions of immense force, caused by the generation of a vast volume of toxic gases, including CO^2 as well as thousands of gigatonnes (1 thousand x million) of poisonous halo-carbons, like *methyl chloride,* arising from the baking of underlying shales and evaporites. Such polluting emissions would have depleted the *'ozone shield'*, resulting in the obliteration of plant-life by exposure to UV radiation, and destruction of the forests. The domino-effect, by cutting off the cycle of transpiration, would have changed the *'hydrological balance'* over the entire Earth, leading to a rapid process of *'desertification'*.

*Y*et, the astonishing mystery is, that our potentially unstable disordered Earth world, appears to have been maintained in a state of *'homeostatic order'*, regulated and kept in order, as a home of exquisite beauty- just right to support *Life Processes,* whilst we know it could be reduced in a trice **to the state of a waste-land**. (**'*Fracking Hell: Earth's weak spots: Hit it in the right place and all hell breaks loose: 255 million years ago, our planet boasted lush forests packed with 30 metres tall relatives of the tiny club mosses and horse-tails that exist today. Some 252 mya, this world was gone – transformed by the most traumatic wave of extinctions this planet has ever experienced—which wiped out all life forms –even the insects suffered, a hardy group which generally sails through extinctions unscathed. Microbes took over,* *and it took 10 million years more to wrest back control'):* Colin Barras: NS.14 Dec.2013. **(88).**

Therefore, the enigmatic inference can only be that, although we know our planetary world to be existing in a state of ephemeral impermanence, and *Life forms* to be in a transient state of reality, the emergence of *Life* can only be explained as emanating from, and imbued with, a creative sense of order. Scientists and Philosophers alike, have found it confusing to discover that the concept of *'solidity & a material world'*, may be illusory, and that what was regarded as real in geographical experience, may not be the true reality. However, if this opening of understanding is considered as *'a phenomenal awakening of spirituality in the human mind'*, we may begin to view ourselves and the world in a different way, with a new sense of wonder, as well as a new set of values: the values of quality, instead of measures based on acquisition and profit, which have come to dominate world systems of government and human organization. In this light, knowing that it makes sense to take care of our delicately balanced home-world. By acting as custodians rather than grabbers, we may also cultivate a *'cybernetic sense of wisdom'*, recognizing the **unity of knowledge, geographical experience, perception of the world, and cultural development.** *('Deep Ecology recognizes the intrinsic value of all living things: Nature and the Self are one--Ultimately ecological awareness is spiritual awareness'):* Capra, F. &Luisi, P.L.:'A Systems View of Life: A Unifying Vision'.

That outlook also defines the *'Discipline of Geography: Queen of the Sciences'*, charged with the specific duty of monitoring interactions linking geophysical world systems with human perceptive behaviour, and systems of organization. *('For Geographers, problems such as water pollution and population bring together all the elements of the subject: historical and anticipated changes; planetary change; local impacts; geo-morphological processes;*

tectonic adaptation; hydrological and climatic dynamics; all wrapped up in a socio-economic context—within a political framework'): Danny Dorling & C. Lee: *'Geography':* Ch.4 *'Sustainability': pp.93-120.*

It would be salutary to keep in mind that all geophysical systems are subject to fluctuations which can lead to a greater magnitude of change than we may expect, and frequently very different from what was anticipated. Whilst scientists and politicians are preoccupied with the impending crisis of *'climate warming',* there is always the probability that the outcome from what we are observing could be surprisingly different from what is projected, because of the complex interplay of feedback loops, to which all systems are prone. Thus, a cogent argument has been made, **'Climate warming will end, not by Fire, but by Ice':** *Felix, R.W.* **(89).**

Principle of Universal Order: Maintenance of homeostasis: The real challenge confronting world governments is to make pragmatic use of the valid knowledge we possess, at every scale of endeavour. For example, we have known for decades that it would make good sense for traffic to drive on the left in *"mid-latitude climatic regimes"* in the N. hemisphere, because driving on the right, as customary in the USA, can act as a vortical trigger to set off the destructive hurricanes which have turned the prairie States into *'dust-bowls'.*

The spin effect imposed by traffic driving on the left, would act as a brake to the formation of violent hurricanes. At a planetary scale of management, knowing how fragile our *'atmospheric system'* is, it would seem sheer madness to subject it to destructive bombardment, by the thundering turbulent effects of space rockets. Also, knowing that the contrails associated with increasing air transport, have had a significant deleterious effect upon the *'Earth's albedo',* which acts as a regulatory cooling system, the current rate of pollution resulting from aeroplane emissions cannot be sustained. (*'Our future is looking less cloudy-and that's far from good news-without them the Sun would obliterate life'*): *NS. 6 Sept. 2014: 2983: Stephen Battersby: 'Clearing Skies'.* (**90**).

Future air-traffic will probably take the form of *'Electrically propelled dirigibles'*, travelling at about 300 m p h, made of a tough new material formed from banana leaves.

Health hazards: 'Covid 19' Pandemic: We also have learned that it is within our grasp to make use of the knowledge gained concerning the health effects of harmful emissions, in relation to patterns of circulation, to maintain air quality in a vital state of purity. How therefore, can crazy behaviour be tolerated, such as *'Formula 1 Racing'; 'founding a mining colony on the moon'; 'sending human beings to Mars'; 'addicting the younger generation to war-games by offering rewards for skill in destroying the enemy'*, etc.? Certainly, profligate, and pointless ventures of this kind do not reflect an attitude of respect for *'Life'*, but rather a morbid and abnormal fixation on annihilation, recognized as,' *the suicidal syndrome of running the gauntlet with death'*. They also reflect a deadening mood of pessimism, which comes from a repugnant view of the world, as an unpleasant and disorderly place to live in, inhabited by hostile people. No wonder that so many of our youth are exhibiting signs of abnormal withdrawal from what may appear to them as a place full of danger. Nothing could be further from the truth. Geographically based understanding of *'Life in a world-environment'*, equipped as wonderfully interlinked and intricately enmeshed organized systems, reveals fanciful and futile speculations about the probability of intelligent life being widespread throughout the Universe, as baseless nonsensical assumptions.

Erroneous and divisive concepts: St. Paul, as he was known by his Roman citizenship, must have seen it a profound mistake to regard disastrous events as punishments, inflicted by the *'Will of God'* for the sins and wicked behaviour of man. Christ's words were explicit: God, our Father, does not condemn what He, or She, has brought into being. However, he could not have been fully aware of Christ's teachings, and all He did, to make known and explain how much, *'God loved the world and all He had created':* *('Yes, God loved the world so much that he gave His only son, so that everyone who believes in Him may not be lost but may have eternal life. For God sent His son into the world, not to condemn the world, but so that through Him the world might be saved. No one who believes in Him will be condemned, but whoever refuses to believe is condemned already'):* St. John: 3. 16-21). **(91).**

The four synoptic Gospels, we have come to rely upon, to learn and know *'the Word and Will of God'*, were not written before St. Paul became converted by his *Damascene* experience, in about 40 AD. *(Gk. Paul= name meaning determination, purpose: Hebrew: Saul= asking, inquiring).* He would have heard all about the Crucifixion, and the wondrous events accompanying the death of Christ, His Resurrection three days later, and how the Roman Centurion (Longinus), was ordered to ensure that Christ was dead, by piercing His side and heart with a lance, and the two thieves who were crucified with Him, by breaking their legs. He would have heard all the gossip concerning these events, spoken in all the languages and dialects throughout the Mediterranean world, with which he was familiar. He would have recognized that Longinus had experienced the same flash of enlightenment as he had experienced himself, when he saw blood and water issue from the wounded heart of Christ, causing him to exclaim, *'He was truly the Son of*

God', and to become a devoted member of *'The Assembly of Christ'*.

St. Paul would also have been inspired by St. Peter's address to the people in the Temple, which brought the faithful together as *'The Christian Body, followers of Jesus Christ'*: *('Now I know brothers, that neither you nor your leaders had any idea what you were really doing: this was the day that God carried out what he had foretold, when he said through all His prophets that His Christ would suffer. Now you must repent and turn to God, so that your sins may be wiped out, and so that the Lord may send the time of comfort'):* Acts of The Apostles: 3. 17-21.

('The whole group of believers was united, heart and soul, no one claimed for his own use anything that he had, as everything they owned was held in common. The apostles continued to testify to the resurrection of the Lord Jesus with great power, and they were all given great respect. None of their members was ever in want , as all those who owned land and houses would sell them, and bring the money from them, to present it to the apostles; it was then distributed to any members who might be in need'): Acts: 4. 32-33. **(92).**

He would also have recognized the same voice of Christ, who had called to him, *'Saul, Saul, why do you persecute me?'*, in Christ's last words spoken on the Cross, *'Father, forgive them for they know not what they do!'*. Thus, together with St. Peter, St. Paul contributed to the evangelization of the old Phoenician city states, by setting out on his onerous missionary journeys, to gather the *'Christian assembly'* together, who then affiliated with the *'Patriarchate of Antioch'*. Antioch, the ancient Greek city on the east of the River Orontes, became the capital city of the Roman Province of Syria, and the 3rd largest city of the Roman Empire, now in ruins near Antakya in Turkey.

Saul was enabled in a flash of enlightenment, to see that the bond which unites and overcomes all hateful divisive temptations, was a higher cognitive spiritual charisma of love, compassionate and actively available to every human mind, inspired by faith in *'The Word and presence of God'*. He would also know himself, and review his life with critical insight, mortified on knowing for the first time the harm he had inflicted on others, by his former hateful and murderous behaviour, as a Jewish zealot. In that pain of self-judgement, he was made to confront the hideous effects of suffering and disorder, inflicted upon Life and world order, as a result of extremist hateful behaviour. Thus, St. Paul became a dedicated apostle, determined to spread Christ's message, emphasizing the unifying healing power of *Faith & Love.*

Unity of Knowledge, Biogeographical Experience and Perception: Similarly, just as St. Paul contemplated all that was happening in the world around him, collapsing into a state of ruinous disorder, some 40 years after *'The Passion and Crucifixion of Jesus'*, John Henry Newman and Bishop Edward Pusey (Anglican Theologians), appalled by the subcultural decline in religious faith in *'The Church of England'*, together with other committed Anglicans, founded *'The Oxford Movement'*, with the aim of restoring the *'Catholic ethos'* (Universal faith in Christ), among the dwindling *'Christian Assembly in their Liturgical practice'*. In the same way that St. Paul's mind was opened, they saw that what people came to believe, was influenced, and coloured by experience and perception of other people's behaviour, as well as the quality of living conditions in the world as they came to know it.

At that time, the way of life for the working population in *'The Satanic Mills and Factories'*, during the early 19th Century in England, forced to live in hastily built slums, amid cramped unhygienic conditions, was grim indeed. Disease was rampant; infant mortality was soaring; society was breaking down, torn with bitter resentment, inflamed by the lingering effects of festering psychotic wounds inflicted by sufferings in the distant past. The European World appeared to be collapsing into a chaotic state of extreme disorder, again tottering on the brink of all-out war. It cannot be dismissed as simply a coincidence, confronted as we are in the 21st century with the prospect of our world and systems of organization, on the point of impending disintegration and chaotic disorder, that Christ's message is sounding in our ears, loud and clear: **amplified with even greater insistence by scientific discoveries, converging with educative theological teaching**.

(*'The ability of the eye to see consistent colours and forms in a Universe that physicists know to be a shifting quantum kaleidoscope: The strange swirling patterns forming in clouds- filtering the light and reflecting it -until the whole sky (atmospheric constituents+ chemistry) appears as a choreographed spectacle- a side of nature that mainstream Physics had passed by- at once fuzzy and detailed, structured and unpredictable. There is a surprising order in the chaos that develops in the human heart: Ecologists have explored the rise and fall of 'gypsy moth' & bee populations: Economists have tried new theories of analysis. The insights that have emerged have led directly into the 'natural' (environmental) world- the shapes of clouds, the paths of lightening, the microscopic intertwining of blood-vessels, the galactic clustering of stars, etc., etc.,): James Gleick: 'Chaos: Making a New Science': 1987.* (**93**).

With the same insight, as deduced by St. John Henry Newman, and reiterated by Pope Francis (Scientist and leading Theologian & Theosophist), nothing makes sense, considered in isolation. All that we come to know, believe and perceive can only be comprehended as a '***Unity of Creation***'. In this light, it would be well if this message were taken to heart, to heal, as well as to forgive and forget divisive and erroneous concepts and grievances, smouldering on from the past. In that way, it would be good ecological sense to make pragmatic use of the knowledge we have learned, acting as custodians aware of *'The wonder of being and living in a lovely world'*. *('Two features of Newman's approach remain particularly significant: first, there is a unity that binds all knowledge together; and second, that unity involves a unifying thread, relationships – an interconnectedness which may indeed be called wisdom: "He was affirming the integrity of two disciplines – intellectual and religious, going on at once side by side, by a sort of division of labour, and only accidentally brought together"):* Prof. Roderick Strange : 'Newman: The Heart of Holiness': Ch.9: 'A Talent for Educating':pp.124-129.* **(94).**

Fig. 10: Phoenician & Ancient Akkadian Hebraic/Arabic Language: Etymological origin of words & origin of Indo-European Languages. Was this *'Euphrates', Land of Good Brotherhood?*

Fig. 11: Gilgamesh Stone: Vague Record of Catastrophic Great Flood: Probably associated with a crisis of climate change, and the mass migration of whole populations.

Fig. 12: Elegant Minoan Ladies: Palace of Knossos in Crete: Evidence of a highly cultured and sophisticated civilization, a *'Matronal Society',* when womanhood was held in reverence and women were regarded with great respect.

Fi0g.13: Phoenician Commercial and Trading Network: Wealth and economic development underpinning the Civilizations of the Mediterranean Region.

Fig. 14: The People of Ham, second son of Noah: Recorded as having populated Africa and parts of Asia (Hamites from Hebrew word). Egypt is mentioned in The Bible as *'Land of Ham': Psalm 78. 51. ('Related to People of Shem (Sem first son of Noah but divided by conflict and competition').* There is a mystery concerning Noah & *'The Curse of Ham':* *Genesis:9.18-27.* (**95**).

Fig. 15: The Descendants of Shem: Sem first son of Noah: Known as *'Semites (Semitic), from Hebrew word "neshemah= breath".* **Note:** *Gk. 'psukhe'* = breath, life, soul, mind. Following the *'Great Diaspora'* (mass migration of the Jewish people), it is highly probable that all people speaking a language derived from the ancient *'Akkadian (Assyro-Babylonian-Hebraic' base language: "Indo-European" group of languages,* may be regarded as related. We may be able to see ourselves as **one great family! Why are we therefore always at war with each other?***(Gk.: homoios= similar, like + stasis= standing still, = in a state of equilibrium, homeostasis =dynamicaly stable system regulated by regulatory processes that counter external disturbance: enigma from ainos= fable, ainissesthai=speak allusively, obscurely: ainigma= =perplexing mystery:*

murios=countless, innumerable: magma from massein=to knead, make malleable, molten:

phreatic from phrear=well, water below the water table, change phase to steam:

katastrophe from katastrephein from strephein=to turn +Kata= worse=disastrous misfortune:

kubernetes=steersman, control – science of control and guidance, Cybernetics:

energetikos=active from energein= operate with powerful effect:

diasperein from sperein= sow, scatter- diaspora=scattering, mass migration of population:

pragmatic from pratein=apply- to act with comonsense, put to good use:

leitourgia from leos=the common man,people+ergos=working= act of public worship,Holy Mass:

etumos=true, origin: etumologia= origin, etymology of words and language:

euaggelion from aggelein=to announce+eu=good= to bring good news:

Theosophos from sophos=wisdom= wise belief concerning God and Creation:

ephemera from hemera=day =short lived, lasting for a very short time, transient.

Jerusalem from Hebrew- yeru=to flow+shalem=complete and whole=the way to complete Unity.

Musterion from mustikos=secret, =secret knowledge or ceremony.

'What have we learned from our ever increasing knowledge of ourselves, and things of the world?'

In consideration of this question, I have to keep in mind that my perception of the world has been strongly influenced by my experience as a lifelong student and teacher, shaped by the *'Disciplines of Geography and Education',* and preoccupied with the mystery of how *'Life and living things came to be in such a beautifully arranged world'.* I soon found that all sciences have been engaged in gathering knowledge about the physical attributes (systems) of our intriguing home-world, as well as studying the symbiotic interplay and relationships between living things and life processes (habitats), linked with locational human behaviour, (systems of organization). This **'learning process'** has gathered impetus, emerging as a phenomenal enlightenment of the **human mind,** to know our world in a very different way, focused upon the impact of human activity upon the world's fragile **bio-spherical systems.** *('Ever since the elucidation of the molecular basis of living systems, we have known that all elementary processes of life are governed by information. Thus, information turns out to be a key concept in understanding living matter (Kuppers: 1900). More than that: the flow of information at all levels of the living system reveals the properties of communication. This means that the information stored in the genome of the organism, is expressed in innumerable feedback loops—a process through which the genetic information is continually re-evaluated by permanent interactions with the physical environment to which it is exposed. In this way, the living organism is built up step by step, into a hierarchically organized network of unmatched complexity-- The fact that all phenomena of Life are based upon information and communication is indeed the principal characteristic of living matter'):* Bernard Olaf Kuppers: *'Information &*

Reolity':Ch.9: 'Information & Communication in living matter': pp.217-236. **(96).**

In this context, leading Theoretical Physicists, concerned with the strange secrets of *Quantum Mechanics,* continue to express bewilderment over the discovery that '*Matter',* and all the attributes of the world can no longer be regarded as material forms, and therefore the concept of a '*material world, consisting of solid things',* must now be regarded as illusory. *('Isolated material particles are abstractions, their properties being definable and observable only through interactions with other systems—subatomic particles are not things, and so on. In quantum theory we <u>never end up with things; we always deal with interconnections-- This is how new Physics reveals the "oneness of the Universe"--we</u> cannot decompose the world into independently smallest units—As we penetrate into matter, we do not perceive any isolated building blocks, but rather a complex web of relations between the various parts of a unified whole'):* Capra, F. & Luisi, P.L.:'The Systems View of life': Ch.11: 4:'From the parts to the whole': pp. 63-83. **(97).**

James Maxwell (1831-1879), in his pioneer study of '*The Electro-magnetic Field',* had demonstrated that '*Light' was electromagnetic radiation',* which transmitted the mysterious force of '*Electricity'.* That discovery alone has exerted a drastic effect upon the world, acting as a violent chain-reaction of change, for the better as well as for the worse. I can remember that we still relied upon gaslight in the 1920's, until my father took the initiative to have the house connected to the main electric cable, which had only been installed in the main road at some distance, during the previous decade. I recall that the neighbours were reluctant to share the cost of having a trench dug for more than a hundred yards, to make the connection to what was then known as DC Electricity Supply. Within a few years this had to be changed to AC (alternating current), because of the risk of electrocution, involved with a DC supply, Whilst living conditions for the relatively well-off have improved dramatically throughout the world, as a result of the benefits stemming from electric light and electrically driven gadgets, there is a downside: the majority of people, living in poverty in underdeveloped regions, do not have access to these benefits, and become poorer as a result. Also, advanced nations continue to unleash the power of *'Electric Force'*, to produce even more lethal weapons and vehicles of war. Thus, our beautiful and remarkably ordered world becomes a more dangerous place, and the downtrodden become ever more aggrieved and resentful, pessimistic and fearful of the future.

('The actions of our physical world at the quantum level are indeed very counterintuitive, and in many ways quite different from the classical behaviour that seems to operate at the more familiar level of our experiences. The quantum behaviour of our world certainly includes entanglement effects over many metres, at least as they only involve quantum level objects, such as electrons, photons, atoms, or molecules-- can it really be the case that there are two kinds of physical law, one of which operates at one level of phenomena and the other of which operates at another – from the time of the ancient Greeks and before, it had been believed that there must be one set of laws for the heavens, and a completely separate set holding here on Earth'): Sir Roger Penrose: Distinguished Mathematician & Quantum Theorist: Nobel Award 2020: 'Shadows of the Mind':Ch.6:'Quantum Theory & Reality'. **(98).**

'Maxwell's Demon', **or, could it have been,** *'Maxwell's Angel'?* : With the light of hindsight, James Maxwell may well have glimpsed the strange *'quantum behaviour'* of the interactions he was observing in the world around him, as he was contemplating he transmission of *'Energy in the Electro-magnetic Field'*, which enabled him to see all the attributes of the world as a connected whole. Had his mind been opened to

understand, *'light, the mysterious force of electricity'*, including himself and the developing knowledge he had gained, *'as a reality of entanglements'*. Certainly, Maxwell's message continues to unfold with compelling intensity in the turbulent events overshadowing our world in the 21st Century, related to control of *'Energy fields of flow'*.

We may be reminded of the *'Four Angels of God': Michael, the Archangel close to the 'Mind and Will of God'; Raphael, the Archangel of Healing & Light; Gabriel, Archangel renowned as the Messenger; & Uriel, Archangel of Wisdom, guide in discernment of what is good or bad: Hebrews 12. 22-29: Colossians 1.16-17: Psalms 148. 2-5: Nehemiah9-6: Quran 112: 1-4.* **(99).**

Discontinuity between 'Knowledge' & Human behaviour: Why is it apparently so difficult to apply the knowledge gained by our most brilliant scholars to wise use, for the good of our home-world, and the well-being of all people? This is the most pressing question for Geography students to consider, charged with the discipline of monitoring the interactions between world systems. Consider for example these anomalous announcements: *('NASA: The robotic probe recently landed on Mars, has succeeded in producing oxygen from the atmosphere consisting of carbon dioxide': & 'Covid 19' Pandemic': Crisis in India; thousands dying for lack of oxygen'): BBC World News: 26 April 2021.*

Plato posed the same question to the *'Athenian World': 'The world is full of error, to be resolved'*. For that reason, he inaugurated his *'Academy'*, with the aim of gathering valid knowledge. James Maxwell, 2,000 years later, puzzled about the probability of *'Two Electric Forces'*, one slower than the other, as **a counteractive influence, which could provide a cooling effect, acting logarithmically** (in inverse proportion to the hotter input), **in order to check the exponential** tendency towards chaotic disorder, and preserve a state of systemic 'homeostasis'. (100).

The Scale Problem for the Human Mind: That observation draws attention to the high probability that all we perceive to have come into existence, bears evidence of an innate and over-arching **'Principle of Order' in all world systems, for the teleonomic purpose of preventing exponential turbulence rising to a state of absolute chaotic disorder.** Not only does this call into question, Newton's 2nd Law of Thermodynamics, predicting that Energy in the Universe (including life and our world), is destined to dissipate to a state of **Entropy,** absolute and complete disorder, but also raises the problem of comprehending the complex web of interactions involved, as **'The Realities of Geographical Experience'.** *('What role do we have as human beings who perceive, make decisions, laugh, cry in this great fresco of the world, as depicted by modern Physics--if the world is a swarm of ephemeral quanta of space and matter- then, what are we? We are nodes in a network of exchanges—but we are also an integral part of the world which we perceive, not external observers, but made of the same atoms and light signals as are exchanged between pine trees in the mountains and stars. Here, on the edge of what we know, in contact with the ocean of unknown things, <u>shines the mystery and beauty of the world'</u>): Carl Rovelli: 'Seven Brief Lessons on Physics': Ch.7: 'In Closing: Ourselves'.* **(101).**

Professor Rovelli's little *'Book of Lessons'* provides a brilliant succinct summary of the chain of incremental scientific knowledge, leading to the startling revelations of *'**Quantum Theory'** (QED: Quantum Electrodynamics), proving that the concept of *'matter',* and the perception of *'Life and living in a material world',* must be regarded as illusory. We now know that *'nothing can be substantiated as being in a state of solidity',* and that we ourselves are constituent parts of a great Life-world system, within a greater phase state of Universal

Order, far beyond human understanding. Yet, somehow-as Einstein remarked, we are able to be aware of this infinite chain of entanglements. That observation, like all Einstein's *'mind-experimental meditations'*, proclaimed a profound truth, **the unity of knowledge–> probabilities emerging in the form of matter–> learning of the human mind–> unfolding realities of bio-geographical existence: 'Que sera, sera!'; 'whatever we think, say and do, ripples on to colour all we experience anew!'.**

That appears to be the same message received from 'Quantum Theory': **all phenomena (facts, events, perceptions, beliefs & behaviour), can be understood as space-time fluctuations regulated by dominant energy gradients.** We may learn from this sequential accretion of valid Knowledge, that the human mind is being opened in a very intense and special way, at this time in the 21st Century, drawing information gathered by the most brilliant scientific minds in the past, together with the teachings of distinguished Theologians and Philosophers-both past and present, in a dialogue focused on the mystery of **being alive in a beautiful and strangely ordered home-world containing all we need. (102).**

Outdated scientific assumptions based on misleading concepts of, *'a material world',* and *'the origin of Life as just an haphazard event',* can no longer be upheld.

('Processes in an atom may take a millionth of a billionth of a second to be completed; within the central nucleus of each atom, events are even faster. The complex processes that transform an embryo into blood, bone and flesh involve a succession of cell divisions, coupled with differentiation, each involving thousands of intricately regroupings and replications of molecules. This activity never ceases as long as we eat and breathe, and our life is just one generation in humankind's evolution, an episode that is itself just one

stage in the emergence of the totality of life—the emergence of human life here on Earth has taken 4.5 billion years— earlier stars must have transmuted pristine hydrogen into carbon, oxygen and other atoms—this has taken about 10 billion years – so the present visible Universe must be about 10 billion light years across-- a startling conclusion – the very hugeness of our universe, which seems at first to signify how unimportant we are in the cosmic scene, is actually entailed by our existence'): Martin Rees: 'Just 6 Numbers':Ch.1:'The Cosmos & the Microworld':pp.1-10. **(103).**

Refer also to Fig. 5: 'Quantum Dynamics'. Every human being (unique macro-molecular assembly), that ever was or will be, is meant to be here as an essential part of a greater **'System of Creation'.** Theologians and Philosophers have expressed these thoughts in various ways, from time immemorial, stressing that every person and living thing has a contributory part to play in this **'Wonder of Creation'.** Scientists now acknowledge this *'Information Age',* compelled to review Life and the physical world as intricately ordered interacting systems. Rovelli's *'Lessons'* demonstrate that the human mind is endowed with an extraordinary capacity to **'learn more from what has been learned'.** *('it is part of our nature to love and be honest—to long to know more and continue to learn. Our knowledge of the world continues to grow'):* '7 Brief Lessons': p.78.

This stream of epistemological knowledge has opened our understanding of the astonishing and breathtakingly beautiful Earth-world, in which it is our good fortune to be alive:- **Carl Gauss (1777-1855): "Greatest Mathematician since Antiquity"-> A way of formulating the curvature of surface, showing that no completely accurate flat map of the Earth can be made, & studied Earth's magnetic field->'Theory of Energy of Matter'--> James Maxwell->'Theory of Heat'- puzzled that the dissipation of energy to a state of Entropy,** *complete disorder ("Newton's 2nd Law of Thermodynamics"), could be countered by a 'cool electric' force'--> Rutherford/Bohr (1913): 'Spin-Model of atomic systems, "Dense nucleus surrounded by shells of orbiting electrons":-->Bernhard Riemann (1826-1866):'Properties of a curved space, laid the foundation for Einstein's " Theory of Relativity: → Roger Penrose (2005); "The Unity of Knowledge", 'Matter is an illusory concept: Penrose/Riemann "Spin Theory"->'Space, Time, & Energy -> packets of light/photons/energy—quanta, abstract until emergence in the form of material systems'. (104).*

If *'Matter, Space, & Time are all forms of Energy on Earth'*, and the perception of the attributes of *'Life in our Earth world'*, as material, solid entities -subject to fixed *'Laws of Thermodynamics'*, proves to be illusory, speculative assumptions about, *'Black Holes, Remote Galaxies, History of Time' etc.)*, must certainly be called into question: e.g., *('Massive collision of 2 Black Holes, which occurred 7 billion years ago, has only just reached us'?: Princeton Un.: BBC World News: 2020.*

But were we not told by leading Physicists that not even light could escape from Black Holes? So how can we see them now? Wisdom has something to tell us about this: *(' May God grant me to speak as He would wish, and express thoughts worthy of His gifts, since He is Himself the guide of Wisdom, since He directs the sages. We are indeed in His hands, <u>we ourselves and our words, with all our understanding, too, and technical knowledge. It was He who gave me true knowledge of all that is; who taught me the structure of the world and the properties of the elements, the beginning and the middle of times, the alternation of the solstices, and the succession of the seasons; the revolution of the year, and the position of the stars; the nature of animals and the instincts of wild beasts; the powers of spirits; and the mental processes of men; the varieties of</u>*

plants, and the medical properties of roots. All that is
hidden, all that is plain, we have come to know, instructed
by Wisdom who designed them all'): Wisdom: 7. 15-22.
(105).

These didactic thoughts must surely be regarded as the
distillation of human experience, rather than promptings
based on assumptions and phantasies. Whether one believes
in a _'Creative God',_ or not; or whether your inclination is to
devise some kind of mental interface, such as _'Nature has_
thought of this and has some tricks up her sleeve': the point is
that discoveries such as, _('E= R curvature of earth':_ _Gauss:_
& E=mc²:Einstein), show that advances in knowledge
emanate from a dialogue with a greater mind, because we
know that it is not possible for the human mind to grasp the
incredible network of complex interactions by which, _**'Life**_
**and life processes have come into being, supported by**
**remarkably propitious Earth-world physical systems'.**

Moreover, the phenomenal _revelations_ of Quantum
Electrodynamics show that the intricately linked molecular
systems, apparently communicating with each other, to make
it possible for the emergence of Life systems, must have been
'fine- tuned' from the beginning. Furthermore, this advance in
knowledge has inspired the ability to assess the knowledge of
one system, by comparison with another, in terms of the ratio
of _**'Energy flow gradients':**_ _e.g._ _**'Maxwells Paradox',**_
showing that all world systems appear to be regulated
according to **'a principle of Universal Order',** **acting as a**
corrective immunological response to counter the
probability of _'turbulence cascading into chaotic disorder'._

Christ made this very clear, in words that a child can
understand: _('You know how to read the face of the sky, but_
you cannot read the signs of the times—He put this question
to his disciples; "Who do you say I am?" then Simon Peter
spoke up: "You are the Christ, Son of the living God". Jesus

replied, "Simon, son of Jonah, "It was not flesh and blood that revealed this to you, but my Father in heaven: So I now say unto you- you are Peter, and upon this rock (firm foundation), *I now build my Church* (Assembly of Judaic-Christian people)-*And the gates of the underworld can never hold out against it"*: Mathew 16. 1-20.

It would be foolhardy to disregard sound advice, expressed by wiser and greater minds than now exist, at this time when the world appears once again to be in a mess. Why are scientists, especially Physicists, apparently so averse to acknowledging the teachings of Christ, known to all who heard Him as *'The Healer'*, and who left no doubt that *'Reality is not what we think it is'*.

In that light, we may understand that our world, and every constituent of it, including thoughts expressed in meaningful language, as well as deeds inspired by the store of goodness in human hearts, has the potential to transform our home-world. *('Then He said to His disciples, "Do not worry about your life and what you are to eat, nor about your bodies and how you are to clothe it- For life means more than food, and the body more than clothing. Think of the ravens; they do not sow or reap; yet God feeds them. And how much more are you worth than the birds—if the smallest things, therefore, are outside your control, why worry about the rest—Your Father well knows what you need. No; set your hearts on his kingdom, and these other things will be given you as well')*: Luke. 12. 22-32. **(106).**

However, epistemic advance in knowledge depends on keeping an open mind, **in a world full of error,** susceptible to negative influences, such as **atheism,** which tends to generate a pessimistic outlook, devoid of hope towards what may happen in future, and is conducive to mental, moral, and cultural decline. Indeed, we are witnessing this dark influence, clouding our modern world like a creeping contagious mental malaise, in the form of *'virtual world experience'*, promoting addiction to *'gaming' & gambling'*, particularly among the young generation, with the lure of *'getting rich quickly & enjoying yourself while you can'*.

('Admitting uncertainty (keeping an open mind), *not only bridges the divide between science and religion, but also can do the same when applied to Life's seemingly perpetual cycles of dispute'): Richard Feynman: Pioneer of Quantum Theory.* **(107).**

Loss of scientific knowledge from the past: In 1961, the wreck of an ancient Roman trireme was discovered, which had foundered in relatively shallow water, off the coast of Antikythera in Greece. It was loaded with bronze and other metal objects, plundered from ancient Greek Temples and Shrines, and was on its way to Rome with a cargo for recycling, when it sank during a storm. A box was discovered amongst the bits and pieces, which turned out to be the oldest known analogue computer, dating to 150-100 BC, a device to predict astronomical positions and eclipses decades in advance. It was traced to the studies of Hipparchus of Rhodes, who had used some device to track the 4-year cycle of the *'Olympiad'* (Olympic Games).

These events were regarded in ancient Greece as a *'Chivalrous Code of Honour'*, requiring all disputes and wars to be suspended and replaced by sporting events, mutually enjoyed and respected, for the purpose of restoring harmonious relations and good order, lest hostilities should get out of hand and become outrageously destructive. How sensible! The Olympiad Games bear the hallmark of *'feminine empathy'*, endowed in the *'mind of woman'*, especially for the purpose of keeping the family bound together, with love and care for each other, as well as to mitigate the brutish behaviour of men: *(L: religare= bind together:* True religion= Care for each other & well-being of the world environment). I

The Antikythera device was studied in Cardiff University, using X-ray tomography and high-resolution surface scanning, which revealed that it consisted of 37 bronze gears, with scratched inscriptions, which enabled Mathematicians, Mike Edmunds and Tony Freeth, to reveal that it was an astonishingly accurate computing device, for tracking the movements of the sun and the moon, and probably for calculating the positions of the 5 known classical planets. This showed that ancient Greek astronomers must have been well-informed about the irregular orbit of the Moon, knowing that its velocity was higher at its perigee than its apogee, knowledge which only came to light again in the 20th Century, although there is evidence that Byzantine and Islamic Scholars had retained some of this knowledge: e.g *Ulugh Beg: Uzbekistan (1336-1405): Astro-Mathematician.*

These Tables were still in use during the 2nd W W, and I have used them myself, serving as a pathfinder navigator in RAF Bomber Command, using a sextant to read angular declinations, to derive a bearing and fix a position. Professor Michael Edmunds, Astro-physicist Emeritus, was sure that there is more to learn about the astounding breadth of knowledge, concerning the perturbations involved in the irregular movements of *'The Earth-Moon Binary System',* demonstrated in the construction of this computer, and discovered by the sages of Ancient Greece. One can only wonder about the source of such profound knowledge, relating to the vagaries and complex interactions in Earth's geophysical systems, showing understanding of mathematical codes of analysis. Whence could they have derived knowledge relating to the analytical methodology of *'Integral Calculus',* involving extremely complicated and abstruse reasoning, requiring the cognitive ability to think in a connected logical way about the workings of geophysical and technological systems, as expounded by Carl Gauss, 2,000 years later?

They were evidently familiar with the 2 branches of calculus: - *'Differentiation'-how the energy of a system/function changes in relation to its variables; & 'Integration'–a method of giving a reliable assessment/summation of a system/function between 2 values.* In our modern world, Civil and Mechanical Engineers use Integral Calculus to analyse patterns of turbulence in dynamic fluid systems, as well as the probability of stress in pipes designed to carry fluids under pressure (*'Hydraulic Systems'*). Technological developments in electronic devices and gadgets, depend on knowledge derived from the analysis of electro-magnetic wave patterns, by *'Integral Calculus'*, to study and control the flow of electrical energy (electrons). When Professors Freeth and Edmunds, succeeded in reconstructing the *'Antikythera Mechanism'*, they expressed their astonishment at the knowledge and highly advanced cognitive power of reasoning, which it revealed. We are reminded that greater minds have preceded us, and it would be naïve to assume that we know better, than wise and learned minds long ago. It has been suggested that *'Archimedes of Syracuse (287 BC)'*, may have been the inventor: *'Decoding the Ancient Greek Astronomical Calculator': Freeth, T. & Edmunds, M:'Nature (Nov.30 2006) & July 2008).* **(108).**

Bridging the gap between religious belief & scientific assumptions: In many ways, the great scholastic minds of the past, appear to be ahead of us, in their cognitive processes, enlightened by an ability to comprehend themselves and the world as a network of organized interacting systems, rather than the limited mechanistic view still held by many today, of a world beset by haphazard events and experiences, in which it makes sense *'to look after yourself and grab what you can, while it's there for the taking'.* That blinkered view of the world has engendered a selfish and pessimistic outlook towards the future, whereas it is evident that the more comprehensive understanding to see our world- and ourselves as parts of it, *'as an integrated whole, rather than dissociated parts, cultivates an all-pervading sense of ecological awareness, recognizing the fundamental interdependence of all phenomena, and <u>the reality, that as individuals and societies, we are all part of-as well as dependent on the cyclical processes of our natural world'</u>*: Capra, F.: *'The Web of Life: A New Synthesis of Mind & Matter'*. **(109).**

In that context, to ignore or gloss over the pragmatic demonstrations of loving care and healing by **Jesus Christ,** during his didactic mission on Earth, would be tantamount to refuting the revelations of *'Quantum Electrodynamics'*, which also affirm the *'unity of all things, encompassing <u>theological & scientific knowledge, and all we perceive, functioning in accordance with an inherent "principle of order", in dialogue with an omnipotent spiritual mind; "The Mind of God"</u>.* Some may prefer to describe it as the working of *'photo-electric effects, in terms of energy gradients'*: but to deny the existence of God, is equivalent to refusing to believe in what you do not know; an unwarranted reductive view of the, **higher logical faculty of 'Spirituality of Mind', whence emerge the healing qualities of ecological loving care,** desperately needed in our disorderly turbulent world, torn by hateful conflicts. That perverse assumption cannot be upheld as valid scientific knowledge, but should be discarded as misleading information, which has spread the false myth, *'that religion has done more harm than good'*. Nothing could be further from the truth. *(L. religare=to bind together=to make all one): ('Ut omnes unum sint' – 'Let all be united as one'):* Christ's Prayer of healing, on the Cross.

The philosopher/scientists of old appear to have been highly aware of the unity of all that is known, and perceived as unfolding events in a capricious unpredictable world, revealed to us in a continuous dialogue between the human mind and a source of omnipotent truthful knowledge: *'The Word & Will of God'? Sanskrit: karman= action, effect, fate: karma marga= to attain a better world by what you say, behave and do: & karma yoga =striving to achieve perfection in this world, and the next phase of existence by controlling desires and putting self last).* The ancient Greeks, regarded presumptuous, self-centred behaviour as offensive to *'the Gods',* invoking punishment as the cardinal sin of *'hubris'.* Written in Hebrew early in the 2nd Century BC., 'The Book of Ecclesiasticus', consisting of the essential teachings and aphorisms handed down from ancient Hebrew Scriptures, relating to wise management in world affairs, as well as to harmonious and peaceful society relationships, was translated into Greek. This provided valuable advice and guidance for Greek speaking Jews, in the light of *'Worldly Wisdom & Social Prudence',* still sorely needed today. *('Pride is hateful to God and man, and injustice is abhorrent to both- Empire passes from nation to nation, because of injustice, arrogance and money. The beginning of human pride is to desert the Lord, and to turn one's heart away from one's maker'):* Ecclesiasticus: 10. 6-12. **(110).**

Metaphysics: 'The ultimate Science of "Being and Knowing": Aristotle, a pupil of Plato at the *'Academy in Athens' (c360BC),* began a critical study of *'The Theory of Ideal Forms',* based on a wide-ranging analysis of *'natural language, common sense, and knowledge contributed by scientific observations & empirical experiments'.* His thoughts, published in his treatise, *'Ta meta, ta phusika'* (the reason why things are as they are), were centred on 3 main questions: **1. What is existence? 2. What sorts of things exist in the world as we perceive it to be? 3. How can things exist and at the same time undergo the changes we perceive to be taking place in the world about us?** In summation, he poses the key philosophical enigma, which continues to occupy the human mind today: **'How to understand the world in the light of unfolding knowledge, including the mystery of human consciousness, as a** *'Quantum Processor'*, **enabling the human mind to** *'know itself'*, **as but a part of,** *'The complex Web of Life'*. In this context, Jerome Bruner (1915-2016), Educational Psychologist and Codebreaker, had a seminal role, as the pioneer of *'Learning Theory & Development Psychology'*, showing that the human mind has a remarkable capacity, to learn from what has been learned. **(111).**

His detailed studies in the *'analysis of Language as the medium and transmitter of thought'*, opened understanding of the cognitive processes by which meaning *(semantic analysis)*, and significance *(semiotic analysis)*, can be conveyed in terms of subtly nuanced language, demonstrating that the educated use of language is closely linked with cultural development. This, in turn has profoundly influenced the development of **'Cybernetics', a new science based on the potential of the human mind as 'an integrated system of control'.('For a system to be conscious, it must integrate information in such a way that the whole generates more information than the sum of its parts-- integration breeds awareness'):** *Giulio Tononi: Neuroscientist:* **(112).**

Not only are we reminded yet again, that the human mind remains continuously engaged in dialogue with an omniscient font of knowledge, but also the amazing spiritual capacity of **'Mind', to heal the rights, wrongs, and festering wounds that threaten to destroy our beautiful world.**

Renewed interest in the meaning of language has sparked a revival in **'etymological studies',** focused on the development and change in meaning of words, as a code of communication. This has drawn scientists and theologians together in their concern to express thoughts and concepts clearly, in view of the uncertainty created in the aftermath of

'*QED*', which shook scientists to the core, with the startling realisation; '*Reality was different from what they had thaught it was*'. *('The concept of a continuous space and time seems doubtful when quantum effects are taken into account'):Paul Davies: Physicist/Anthropologist/Geographer.*

Richard Feynman, was right to the point, '*Uncertainty can bridge the gap between science and religion*', providing all are prepared to think, say and behave in the light of truthful knowledge, and with care for each other.

('The task is to offer the opportunity for continuing education through life, to provide the materials and resources of education for better living, to help people to become more efficient in their chosen field of work— balanced and enriched in their private lives-- to the end that they may bring their skills and humanity (ecological care), to the service of society as a whole'): Lawrence Quincy Mumford (1903-1982), Distinguished Librarian & Educationist: (President of Library of Congress, 1954-1974).

Theological, philosophical and scientific knowledge could be made to bear as an educational influence of immense power, to solve problems of disorder in world systems, easily resolved in the light of brotherly love. *('Each of us has a special capacity to relate to nature, to our fellow man, to ourselves and to G-d. Of course these relationships are not like watertight compartments which are independent of one another. Our relationship with nature results from our observations and the detailed handiwork that we observe; relationships with our fellow man are based on our interests and life experiences; and our relationship with G-d, which comes from deep inside our being, is a result of self-dialogue, and is nurtured by all the other relationships mentioned above. True dialogue is at the heart of a thinking man's life and demands that each person tries to get to know and understand the person with whom they are conversing.*

When conversing with one's fellow man, words are merely vehicles for communicating, although even in societies where everyone speaks the same language, the exact words can take on a different meaning. Each person adds their own nuance to many of the words they use, which then becomes part of the linguistic heritage.--"G-d's candle is man's soul which reveals the innermost parts of his being"): Rabbi Abraham Skorka:'On Heaven and Earth': A Dialogue between Pope Francis and a Rabbi Biophysicist. **(113).**

This remarkable eye-opening account of a discussion between two leading *'Theosophical'* teachers, emphasizing the capacity of *'The human mind to experience moments of spiritual elevation while alone'.* We become aware of all that unites us, in common with each other, having the *'same persistent questions, but expressed with their various interpretations',* in our desire to understand how we came into being alive. *('Each soul is a reflection of the other-- The Divine Breath* (Lectio Divina), *knows how to unite them, and then form a link with Him that will never weaken').* That concept echoes **Christ's prayerful words:"Ut omnes unum sint"**, as well as the rumination of the *'Quantum Theorist': " when we find the uniting theory of all things, we shall see the 'Mind of God'.*

The Omniscient Mind: *('Without a doubt, experiencing God is dynamic- to use a word that we learn in our mutual study of basic science. However, what do you think we can say to people nowadays when we find the idea of G-d to be so mangled, profaned and diminished in importance?'):* Rabbi Abraham Skorka: Ph.D: 'Chemistry': 'On Heaven & Earth'.

('What every person must be told is to look inside him-herself. Distraction is an interior fracture. It will never leave the person to encounter him/herself for it impedes him from looking into the mirror of the heart. Collecting oneself is the

252

beginning. That is where the dialogue begins. At times, one believes he has the only answer, but that's not the case— seek the experience of entering into the intimacy of your heart, to experience <u>*The Face of God*</u>*-- I tell people not to know God only by hearing:* <u>*The living God is He that you*</u> <u>*may see with your eyes within your heart'*</u>*):* Pope Francis: Ph.D: 'Atmospheric Chemistry': 'On Heaven & Earth': pp. 1-16.

That draws attention to the words of Christ: *('For a man's words flow out of what fills his heart.* <u>*A good man draws*</u> <u>*good things from his store of goodness—a bad man draws*</u> <u>*bad things from his store of badness'*</u>*).*

('When we believe that history starts with us, we stop honouring the elderly. Often, when I am a little down, one of the texts I run to is Deuteronomy, Ch.20, to realize that I am just one more link, that I have to honour those who have preceded me, and that I have to allow myself to be honoured by those who are going to follow. That is one of the strongest actions of old age. "There is a whole plan of God walking with each person, and it started with our ancestors, and it continues with our children"--the old person knows, consciously or unconsciously that we have to leave behind a testament of life—we owe so much to those who have gone before us—the wisdom of the elderly has contributed a lot and we should venerate those upon whose shoulders we stand'): Pope Francis: 'On Heaven & Earth': 'On the Elderly', & how much we owe to the wisdom of the past: (with some abbreviation). **(114).**

That intuitive knowledge, which springs from a spiritual quality of mind, may be taken as an affirmation that all we perceive at present, in world systems, has been shaped by the dynamics operating in past conditions, with the continuing effect of colouring the state of living conditions that may be expected to emerge in future. In that way, what has been gathered in the form of a store of *'goodness or badness'*, *'wisdom or folly'*, *'peace or war'*, *'love or hate'*, will assuredly determine the state of the world, to be handed on to our children, and following generations. Thus, it becomes a duty to think about what each of us can contribute to the well-being of our world-home, rather than concentrate on grabbing all we can, beyond our needs.

Veneration of our forebears breeds a sense of obligation towards our young, as well as to our offspring to come. That generates a sense of self-respect, knowing that those nearest and dearest to us, our family, clan, country et cetera, depend upon us. In essence, we begin to value our being, as a valued link in a chain of charismatic loving care. Love of one's own being becomes a prayer of thanksgiving for the wondrous gift of Life, leaving no room for the deadly *'pandemic of pessimal despair'*, now creeping across the world, afflicting the younger generation particularly, with symptoms of mental and cultural *'malaise'*, even to the point of feeling that Life is not worthwhile.

For the discipline of Geography, charged with monitoring the interactions between systems of Human Organization and the Geophysical dynamics of our Earth-world, there is much to do: harmful erroneous concepts and false information to be corrected, as well as faith to be restored in the wonder of being alive in a world of great beauty, in a Universe of exquisitely ordered mystery.

('Oh Lord, I thank you for the wonder of my being, for it was you who created my innermost self, and put me together in my mother's womb. For all these mysteries we thank you—for the wonder of myself, and for the wonder of all your creation I thank you'): Psalm 139. **(115).**

(Gk: dialogos from dialegesthai= to see, know, understand: =conversation, discourse.

Aphorismus from aphor=concise statement, scientific principle, definition.

metaphuesthai=transformation;(Aristotle: "ta meta ta phusika"): Philosophy of First

Principle of Order; concepts of being, essence, time, space, cause: science of unity of all things and knowledge.

Analusis from luein =look into, unloose = search for reason, find the answer

drastikos fom dran= do, happening = violent change

Theosophos from Theo (Deo)=God + sophos=wisdom= Word and Will of God.

homalos= steady state, even: anomalos= unconformable

perigaeum from perigeion=nearest the planet Earth: apogaion=far from Earth.

Etymology from etumos= true meaning of words and language, (history of change).

Technologia from techne=art, knowledge of mechanical systems, how things work.

pragmatikos from pratein=do, from pragma=deed= demonstration.

scholastikos from skolazein=to be at leisure: to find pleasure in study and research.

Khaos= vast chasm, extreme disorder.

Kacophone: from phone=sound: =ill sounding.

Sialon= saliva, surface scum

rhuthmos from rhein= flow: rhythm.

Kineticos from kinein= to move: heat energy produced by movement.

Kheme= transmution of metals, chemical change.

Epidemia =prevalence of disease: from epi + demos:

demos= people, population: demokratia = rule of the people for the good of the people.

Fg. 10: Phoenician script: Alphabet from 'Akkadian' ancient Hebraic/Arabic language

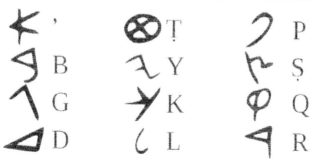

he Phoenician alphabet. Note that ' and ' were originally full
onsonants in the Phoenician language (glottal stop ? and

Fig. 11: Gilgamish Stone Tablet: Record of 'The Great Flood': (British Museum)

□□□□

Fig. 12: Elegant Minoan Ladies:Palace of Knossos, Crete:'*A cultured Matronal Society'*

Fig. 13: Phoenician Commercial and Trading Network

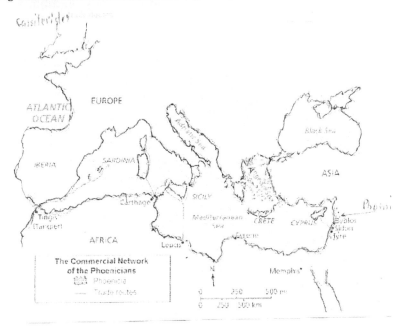

Fig. 14: The People of Ham

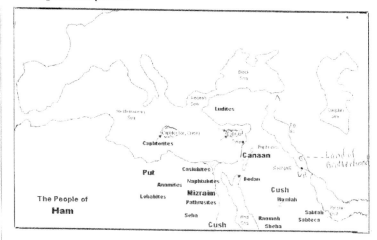

Fig. 15: The Descendants of Shem

The People of Shem

Chronology of events:
Arrival of Sumerians, highly civilized survivors of a vanished nation: c.7000 BC
Akkadian ancient Hebraic language-> Phoenician->base of 'Indo-European languages'.
Sumerian 'Gilgamish Poems': 2700 BC

Noah: b. 2948 d. 2445 BC

'Great Flood', Ark Shem & Ham: c. 2458

'Neo-Assyrian 3rd Dynasty Ur: 2100 BC

Abraham (son of Terah of Ur: b. 2166 BC

Gilgamish Tablet, 'Flood Record': c.700 BC
With acknowedgement to Wickepedia Archives

Elam   Asshur   Arphaxad   Lud   Aram
                   Shelah
                   Eber   Uz Hul Gether Meshech
Joktan   Peleg
                   Reu
                   Serug
                   Nahor
                   Terah
        Nahor   Abraham   Haran
        Ishmael   Isaac   Lot
                   Esau Jacob Moab Ben-Ami
        Ishmaelites Edomites Israelites Moabites Ammonite

260

Epilogue

What is this mysterious bond of Love? How could such an abstract incorporeal sense of awareness, emerge as a force of immense empowerment, capable of healing hateful divisions, replacing subcultural influences with higher spiritual qualities of custodial care, unless there were a *'fundamental principle of universal order'*, commanding *'deep ecological respect'* for the well-being of all living things? That insistent message appears to have permeated the unfolding record of human geographical affairs throughout the ages, mounting to a crescendo at recurring times of impending crisis. **'Mens Sana in corpore sano':** *('You should pray for a healthy mind in a healthy body: Ask for a heart that has no fear of death: And deems length of days the least of Nature's gifts: That can endure any kind of toil: That knows neither wrath nor desire: And thinks the woes and hard labour of Hercules better than the banquets of Sardanapalus* (first King of Assyria): *What I commend to you, you can give yourself: For assuredly the only road to a life of peace is virtue!': 'Satire X': Juvenal: (b.55AD).*

'All things are full of God': *('All that is known and experienced proclaims Nature as the unity of all things': 'Water is the originating principle of Life': Thales (b. 624 BC): Mathematician, Astronomer, & Philosopher: Known as one of the 7 Sages of Greece.* **'There is more wisdom in your body than in your deepest Philosophy':** *Zarathustra (Gk.Zoroaster): Iranian Reformer & Prophet (b. 628 BC): Could he be 'Melchizedek, the Priest of El Elion'?: Genesis: Ch.14.*

'What does a man gain for all his toil: *('Vanity, of vanities. All is vanity: For so it is that a man who has laboured wisely, skilfully, and successfully must leave what is his own to someone who has not toiled for it at all-- what of all his laborious days, his cares of office his restless nights. This too is vanity!'): Book of Ecclesiastes: 1:2; 2-21-23.*

261

'You must look for the things that are in heaven where Christ is': *('Let your thoughts be on heavenly things, not in the things that are on the Earth—put on a new self which will progress towards true knowledge the more it is renewed in the image of its Creator-- In that image there is no room for distinction between Greek and Jew, or between the circumcised and the uncircumcised, or between barbarian and Scythian, slave and free man. There is only Christ: He is everything and He is in everything: St. Paul to Colossian's.*

"**Make up the differences between you instead of disagreeing among yourselves'**: *'For the sake of our Lord Jesus Christ – it is clear that there are serious differences among you-- I appeal to you, be united again my dear brothers in your belief and practice!':St. Paul; First Letter to the Corinthians.* '**United we stand, divided we fall'**: *'They asked Jesus, as a test, for a sign from heaven; but, knowing what they were thinking, He said to them, "Every kingdom divided against itself is heading for ruin, and a household divided against itself collapses": Luke 11.17-18.*

'What is Love'?

Thoughts inspired whilst watching a nurse administer loving care to an elderly patient, in a Nursing Home:-

Easier by far to see where when it's not,

Wondrous fusion from Truth in God's holy mind, Bestowed by our Creator when the Universe was begot, Conferred on the world, to protect and to bind

All together as one, ward off the satanic sword,

Overcome evil, and refine the brutish horde.

"Love is not love, which alters when it alteration finds, Or bends with the remover to remove: Oh no, It is an ever fixed mark," All suffering, forever enduring,

Not taking, selfless, all-giving,

Not hurting, but bent on curing,

With comfort unending, spirit ever living,

Never bruising, but embrace enfolding,

Not dismissing, but ever aspiring,

Never belittling, but rather upholding,

Against despising, always admiring,

Constantly praising, whilst uncomplaining,

Encouraging, rather than criticising,

Ever respecting, never, never disdaining,

Accepting, rather than wound by despising,

Never depressing, but always protecting,

Always desiring, not drawn to dejecting,

Always supporting, with a smile on meeting, Channel of peace with joyful greeting!

Your Maker is Compassion and Love! You believe it's true? Then, why replace love with hate?

Wrap our world in shrouds of woe: Christian-Muslim-Jew?

Sleepers awake! For you may leave it too late,

To reflect God's love: Know that it is required of you!

With acknowledgements to The Bard.

References

NS.=New Scientist : NT =New Testament

1 Bateson, G. :'Steps to a Ecology of Mind': Introduction. Bantam 1997.

2 Tegmark, M. : 'Our Mathematical Universe': Ch.11. 'Is Time an Illusion?': Allen Lane 2014

3 Pross, A.: 'What is Life?': 'How Chemistry becomes Biology': OUP 2012

4 " ": 'What is Life?': Ch.4: 'Stability & Instability': & Ch.8: pp.160-191.

5 " ": 'What is Life?': Ch.1: 'Living things are so very strange!'.

6 Capra F. & Luisi, PF.: 'The Systems View of Life: A Unifying Vision': CUP 2014.

7 Blaise Pascal: Pioneer of Probability Mathematics: 'Pensees'.

8 Davies, James A.: Professor of Experimental Anatomy: 'Life Unfolding': OUP 2014.

9 Capra F. & Luisi,PF: 'The Systems View of Life': Ch.7:'A New Synthesis': pp.153-177.

10 Varela, F.: 'Santiago Theory of Cognition'.

11 Davies, P. & Gregerson, N H.: 'Information & The Nature of Reality': Introduction. CUP 2014.

12 Brooks, M.:'20 Big Questions in Physics':'What is the true Nature of Reality?'. pp. 72-82.

13 Bateson G.: 'Mind & Nature: A Necessary Unity'; Bantam 1979.

14 Chomsky, M.: 'Language & Mind': Brace & World 1978.

15 Lord Martin Rees: Astronomer Royal: 'Just 6 Numbers': CUP 1999.

16 Crick, F.: 'The Astonishing Hypothesis: Search for the Soul': Simon & Schuster 1994.

17 NT: 'Condition of following Christ': *Luke 9. 23-36.*

18 NS: 24 May 2014 :2970: p.20: David Deutsch: 'Why we need to reconstruct the Universe'.

19 Sir Roger Penrose: 'Shadows of the Mind: Ch.6:Quantum Theory & Reality':pp.307-347. OUP 1994

20 Blaise Pascal: *'Knowledge is like a Sphere': Probability Theory'.*

21 Niels Bohr: 'Atomic Theory & QuantumTheory': Nobel Prize: 1927.

22 Heisenberg, W.: 'Quantum Field Theory' : Nobel Prize: 1932.

23 Einstein, A.: *'The Photoelectric Effect': 'Particles & Photons': 1905*

24 Deutsch, David E.: 'The Fabric of Reality': 1997.

25 NT: *'You must do what the Word tells you!': Letter of St. James: 1. 17-27.*

26 Brooks, M.: 'Thirteen things that don't make sense': Ch.3.'Varying Constants': Penguin.

27 Feynman, R: '6 Easy Pieces: Relation of Physics to other Sciences': pp. 47-67.

28 " " : " " : Ch.2: 'Basic Physics': pp. 2-45.

29 Thomas Aquinas (1224-1274): *'Summa Theologica: Quinque Viae'.*

30 Heisenberg, W.: 'The Uncertainty Principle'.

31 Bateson, G.: 'Steps to an Ecology of Mind': 'Conscious Purpose & Nature': pp. 442-494.

32 Wiener, N.: Quoted in 'The Web of Life': Fritjof Capra.

33 Bateson, G.: 'Steps to an Ecology of Mind': 'The roots of Ecological Crisis': pp. 111-121.

34 Feynman, R.: '6 Easy Pieces': Ch. 3; pp.63-65: 'Psychology & Sensation'.

35 Davies, Paul: 'The Mind of God: Search for Ultimate Meaning': Penguin 1993.

36 Sir Roger Penrose: 'Shadows of the Mind':Ch.8.6: 'Physical Phenomenon of consciousness'.

37 Capra, F.: 'The Web of Life: Deep Ecology -A New Paradigm': Harper Collins 1996.

38 Schroder E.: 'Order from Order': The Dublin Lectures, 'What is Life?' 1943.

39 NT: *The Saviour of the World: Water of Eternal Life'. John: 4. 5-22.*

40 NS: 5 July 2014: 2976: Cogdell, R.: 'Photosynthesis: Life Support System'.

41 Feynman, R.:'6 Easy Pieces': Ch. 6: 'Quantum Behaviour': pp.15-138.

42 Capra, F.: 'The Web of Life': 'Synthesis of Mind & Matter'.

43 Capra, F. & Luisi, P.F.: 'Systems View of Life': Ch.17.2.1:'Quantative to Qualitative Growth'.

44 Bishop E. Pusey: *'Holy Eucharist, Comfort to Penitent': Oxford 1843.*

45 Cardinal Newman: *'The Idea of A University': 1852.*

46 NT: Christ's affirmation of the unity of 'Knowledge & 'Lectio Divina': 'Beatitudes': Matt.5.3-15.

47 Staff, H.: 'Information & Reality': Ch.6 : 'Mind & Values'. Editors: Davies, P. & Gregerson N. H.

48 Hughes, R.: 'The Fatal Shore': 1986.

49 Moran, J.J.: 'Know Your World':Ch.3. 'Ethereal Atmosphere' & Ch.4.'The Thin Blue Line'.

50 " " : 'Applied Geography': 'Spirituality of Mind': pp.64-68.

51 Tononi,G: NS: 'Your Conscious Mind': Ch.2:'The Biological Basis of Consciousness'.

52 Holmes, R.: 'Information & Reality'; Pt.3 'Biology': pp. 261-315.

53 Maynard Smith: 'Information & Reality:' 'Concept of information in Biology'.

54 Peacocke, A.: 'Information & Reality': Ch.12: 'The Sciences of Complexity':pp.315-355.

55 Feynman, R.: '6 Easy Pieces in Physics': 'Quantum Theory: Atoms in motion'.

56 Davies, Jamie A.: 'Life Unfolding: How the human body makes itself'.

57 Ward, K.: 'Information & Reality': Ch.13: 'God as the Ultimate Information Principle': pp.357-380.

58 Kings College/Barts Hospital: 'Chinese Study 'Covid 19 Mutations': 2020.

59 Haught, J.: 'Information & Reality':Ch.14:'Information, Theology & Universe':pp.383-400.

60 NT: 'Real wisdom & its opposite': Letter of St. James: 3. 13-18.

61 NT: *'No apparent cause can account for what we see'*: St. Paul: Hebrews 11. 1-3.

62 NT: 'True Knowledge': 2 Peter 1-7.

63 Gregerson, N.H.: 'Information & Reality': Ch.15:'God, Matter & Information': pp. 405-443.

64 Welker, M,: " ": Ch. 'The Spiritual Body' :pp. 444-460.

65 Mon. Roderick Strange: 'Newman, Heart of Holiness': Ch3: 'Watching for Christ': pp.27-39.

66 Mon. R. Strange: 'Newman, Heart of Holiness': Ch.4: 'Life in Christ': pp.41-56.

67 " """ : Ch.7: 'In Darkness': pp. 81-97.

68 " " : 'John Henry Newman: A Mind Alive': 'Seeking Church Unity': pp.93-105.

69 " " : 'Heart of Holiness': Ch. 9: 'Talent for Educating': pp.113-130.

70 Pross, A.: 'What is Life?': pp. 9-11.

71 Rovelli, C,: 'The Order of Time': Ch.6: 'The World is made of events, not things': pp.85-92.

72 Sir Roger Penrose; 'Shadows of the Mind': Ch.6: 'Quantum Theory & Reality': pp.307-346.

74 Lord Martin Rees: 'Just 6 Numbers':Ch.11:'Coincidence,Providence,Multiverse':pp.164-174.

75 NS:17 Aug. 2013: 2930: Alex Petrunin:'Erosion by Earth's Central Heating: p.60.

76 Waltham, D.: 'Lucky Planet: Ch.4: 'Constant Change':pp.47-63.

77 Davies, P.: 'The 5th Miracle: Search for Meaning & Purpose of Life'.

78 Humberto Maturama: 'Santiago Theory of Cognition :1978.

79 Baldwin, M. et al: J. Geophysical Research: DOI:129/97:jD62150:'Oscillations of Vorticity'.

80 Khalili, J.: 'Paradox: 9 Greatest Enigmas in Physics':

Ch.4:'Maxwell's Demon': Black Swan 2012.

81 Brooks, M.: '13 Things that don't make sense': *'alpha'?*

82 Sigurggson, M: !S0R: 'Iceland Geosurvey': 'Volcanic Event': Feb. 2021.

83 Hennigan, W.J: Time Magazine: 13 Feb.2018: 'The New Nuclear Poker Game'.

84 Darwin,: 'Voyage of Beagle':1830-35: 'Coral Reefs': 1842: 'Barnacles': 'Earth Worm': etc.

85 NS: 8 April 2015: 3017:p.37: Jeff Hecht 'How did Earth first acquire its continents?

86 Ward, P D.& Brownlee, D: 'Rare Earth: Why Complex Life is Uncommon in the Universe'.

87 Tohver E. : 'The Great Dying Extinction' : *Chixulub Crater Impact: 65 mya: Cretaceous Period.*

88 NS: 14 Dec.2013:2947:pp.42-46:Colin Barras: 'Palaeogeography: 'Fracking Hell'.

89 Dorling, D. & Lee, C.: 'Geography: Ch.4:'Sustainability': pp. 93-120. Profile Books 2016.

90 Felix, R W.: 'Not by Fire, but by Ice!'

91 NS: 6 Sept. 2014:2983:pp.42-47:Stephen Battersby: 'Clearing Skies, bad News'.

92 NT: *God loved the World and all Creation': John: 3. 16-21.*

93 NT: The Christian Ethos: *'Care for each other and all living things'*: ***Deep Ecology: Discpline of Cybernetics: Biogeographical Science & the New Science of Control:*** Peter 2. 3.

94 Gleick, J.: 'Chaos: Making a New Science': Turbulence leading to Disorder in Systems. 1987.

95 Bible: *'The Curse of Ham'?: Genesis 9. 20-27.*

96 Kuppers, B.: 'Information & Reality': Ch 9:'info.& Communication in living matter';0p217-236.

97 Capra, F. & Luisi, P.L.: 'Systems View of Life':Ch.11:'From parts to the whole':pp.63-83.

98 Sir Roger Penrose: 'Shadows of the Mind': Ch.6: 'Quantum Theory & Reality': pp.307-347.

99 Bible: Cycles of Turbulence: *'The 4 Angels of God'.*

100 Maxwell, James C.: 'Theory of Heat':1872': Refer to Fig. 4;'Reaities linking Life & Earth-world'.

101 Rovelli, C.: 'Seven Brief Lessons on Physics':Ch.7:'In Closing: Ourselves': Penguin 2014.

102 " " " " : Ch.1:'Most Beautiful of Theories': Penguin 2014.

103 Lord Rees,M: '6 Numbers': Ch.1:'The Cosmos & the Microworld': pp.1-10.

104 Sir Roger Penrose:'The Road to Reality':'Matter an illusion': pp.428-430: Vintage Books 2007

105 Bible: Book of Wisdom: 'The appeal to divine inspiration': 7. 15-22.

106 NT: 'Trust in Providence': Luke 12. 22-32.

107 Richard Feynman: 'Bridging the gap between Science & Religion': California Inst. Technology.

108 Freeth, T. & Edmunds, M.: 'Antikythera Mechanism':Nature:444:587-591:30 Nov. 2006.

109 Capra, Fritjof: 'The Web of Life: A New Synthesis of Mind and Matter': Harvard Collins:1997.

110 Bible: Ecclesiasticus: 'Government in Systems of Organization:' Against Pride': 10. 6-11.

111 Bruner, J: 'Acts of Meaning': 1990 : 'The Culture of Education':1996.

112 NS. 'Your Conscious Mind':Ch.2. Giulio Tononi: 'Integration breeds Awareness':pp.16-20.

113 Pope Francis & Rabbi Abraham Skorka: 'On Heaven & Earth': Bloomsbury 2013.

114 Pope Francis : 'On Heaven & Earth': 'Dialogue with the Mind of God': pp. 1-16:

" " : " " " : 'The Elderly & Wisdom of the Past': pp. 96-101.

115 Bible: Psalm 139: 'In praise of 'The Omniscience of God & Creation'.

116 NS: Silk, J: 'Cosmic Conundrums': 8 March 2014: 2959: p.26.

117 Rogers, W.E.: 'Introduction to Electric Fields': McGraw-Hill 954: Quoted by A.D. Moore (Michigan Un.): 'Fields of Flow'.

Notes

Notes

<u>Notes</u>

Notes

Printed in Great Britain
by Amazon